SUSTAINABLE PROTEIN PRODUCTION AND CONSUMPTION:
PIGS OR PEAS?

ENVIRONMENT & POLICY

VOLUME 45

The titles published in this series are listed at the end of this volume.

Sustainable Protein Production and Consumption: Pigs or Peas?

Edited by

Harry Aiking

*Institute for Environmental Studies,
Vrije Universiteit, Amsterdam*

Joop de Boer

*Institute for Environmental Studies,
Vrije Universiteit, Amsterdam*

and

Johan Vereijken

*Agrotechnology & Food Innovations BV,
Wageningen University and Research Centre*

A C.I.P. Catalogue record for this book is available from the Library of Congress.

ISBN-10 1-4020-4062-8 (HB)
ISBN-13 978-1-4020-4062-7 (HB)

Published by Springer,
P.O. Box 17, 3300 AA Dordrecht, The Netherlands.

www.springer.com

Printed on acid-free paper

All Rights Reserved
© 2006 Springer
No part of this work may be reproduced, stored in a retrieval system, or transmitted
in any form or by any means, electronic, mechanical, photocopying, microfilming, recording
or otherwise, without written permission from the Publisher, with the exception
of any material supplied specifically for the purpose of being entered
and executed on a computer system, for exclusive use by the purchaser of the work.

Printed in the Netherlands.

CONTENTS

Preface ix
List of contributors xi
List of abbreviations xv

1. Background, aims and scope 1
   Harry Aiking & Joop de Boer
   1.1. Food, agriculture and sustainability 1
   1.2. The PROFETAS approach 7
   1.3. Origins of diet proteins in Europe (EU-15) 11

2. Environmental sustainability 23
   2.1. Introduction 23
   Harry Aiking & Joop de Boer
   2.2. The protein chains: Pork vs. pea-based NPFs 27
   Harry Aiking, Martine Helms, David Niemeijer & Xueqin Zhu
   2.3. Economic approach to environmental sustainability
        of protein foods 30
   Xueqin Zhu & Ekko C. van Ierland
   2.4. Measuring environmental sustainability of protein foods 34
   Martine Helms
   2.5. Ecological indicators for sustainable food production 40
   David Niemeijer & Rudolf S. de Groot
   2.6. Conclusions 46
   Harry Aiking & Joop de Boer

3. Technological feasibility 51
   3.1. Introduction 51
   Johan M. Vereijken
   3.2. Protein-flavour interactions 58
   Lynn Heng
   3.3. NPF texture formation 62
   Francesca E. O'Kane
   3.4. Designing sustainable plant-protein production systems 67
   Xinyou Yin, Jan Vos & Egbert A. Lantinga
   3.5. Breeding: Modifying the protein composition of peas 73
   *Emmanouil N. Tzitzikas, Jean-Paul Vincken, Krit Raemakers
   & Richard G.F. Visser*
   3.6. Methodology for chain design 79
   Radhika K. Apaiah

3.7. Options for non-protein fractions.. 86
Frank Willemsen
3.8. Conclusions.. 91
Johan M. Vereijken & Martinus A.J.S. van Boekel

4. Social desirability: Consumer aspects.. 99
    4.1. Introduction to consumer behaviour.. 99
    Joop de Boer
    4.2. Socio-cultural potential.. 103
    Joop de Boer
    4.3. Substitution of meat by NPFs: Factors in consumer choice......... 110
    Annet Hoek
    4.4. Substitution of meat by NPFs: Sensory properties
         and contextual factors.. 116
    Hanneke Elzerman
    4.5. Conclusions.. 123
    Joop de Boer

5. Social desirability: National and international context............................ 129
    5.1. Introduction to societal aspects.. 129
    Joop de Boer
    5.2. National policies and politics... 132
    Marike Vijver
    5.3. European and global economic shifts... 138
    Lia van Wesenbeeck & Claudia herok
    5.4. International institutions... 144
    Onno Kuik
    5.5. Conclusions.. 150
    Joop de Boer

6. Emerging options and their implications... 155
    6.1. Towards crop-based solutions.. 155
    Harry Aiking & Joop de Boer
    6.2. Crop options.. 157
    Johan M. Vereijken & Anita Linnemann
    6.3. Combined chains... 166
    *Frank Willemsen, Radhika Apaiah, Dick Stegeman
    & Harry Aiking*
    6.4. Actor commitment.. 175
    Joop de Boer
    6.5. Actor feedback.. 181
    Johan M. Vereijken & Harry Aiking
    6.6. Conclusions... 187
    Harry Aiking, Joop de Boer & Johan M. Vereijken

7. Transition feasibility and implications for stakeholders..................193
Harry Aiking, Joop de Boer & Johan M. Vereijken
    7.1. Introduction...193
    7.2. Environmental sustainability..195
    7.3. Technological feasibility..197
    7.4. Social desirability..200
    7.5. Transition feasibility..204
    7.6. Governmental policy options..207
    7.7. Industrial policy options..209
    7.8. Consumers and other stakeholders...211
    7.9. Conclusions..213

History of the future (epilogue)..217

Index..221

PREFACE

Education and health are important priority areas in modern policy making and research. Health comprises a multiplicity of elements that make up someone's well-being, such as medical care, clean water and air, and good food. The call for healthy food is not only heard in less developed countries, but also obtains a prominent place in a modern industrialized society. But like all good things in life, good food is scarce and increasingly more so. In general, scarcity can be tackled by two strategies, namely by an increase in the volume of the goods needed or by a shift to alternatives that serve human needs more or less in the same way. But are these alternatives of the same quality?

PROFETAS is a multidisciplinary research initiative on sustainable food systems, sponsored among others by the Netherlands Organisation for Scientific Research NWO. It aims to link Protein Foods, Environment, Technology And Society by investigating the possibility of substitution in the food chain with the aim to develop a sustainable food system that is environmentally benign through the use of modern technology which may allow us to produce plant based alternatives to meat based ingredients.

It is clear that such a new perspective does not only call for advanced technological and environmental research, but also for in-depth societal research, as the degree of acceptance of new food systems is critically contingent on perceptions and attitudes of human beings. PROFETAS has made a very important and fascinating endeavour to put the existing body of knowledge together and to develop path-breaking new ideas and unexplored pathways that will help us pave the road to a more sustainable food culture.

In this book on "Pigs or Peas?" the results of a strategic research programme on the opportunities of protein food are thoroughly explored, in terms of sustainability, technological feasibility and social desirability. This multidisciplinary research initiative has not only mapped out bottlenecks and hurdles, but it has also designed a policy toolbox demonstrating that pro-active decision making is far superior to a passive attitude. The joint work of ecologists, economists, political scientists, technologists, biologists and chemists has led to extremely useful insights at the interface of food production, environmental sustainability and societal acceptability. PROFETAS has been able to open up pathways for a major transition in food production and consumption, by not only exploring the food chain, but the entire agricultural system (including biomass for sustainable energy production and sustainable resource use of increasingly scarce freshwater). The emphasis on a sustainable system by reducing the pressure on energy, freshwater and health is a key feature of the present publication, which also contains various useful and operational guidelines for policymakers from both government and industry.

This book is a highlight in multidisciplinary studies on innovative food production and consumption in an era of new scarcity and offers useful guiding principles for a transition towards an ecologically and socially sustainable food system from a global perspective. The authors/researchers ought to be complimented for their creative research results which deserve to be disseminated to a wide audience.

Peter Nijkamp

President NWO

LIST OF CONTRIBUTORS

(* = section authors; see headings for initials; affiliations during the programme)

- **Harry Aiking***, Institute for Environmental Studies, Vrije Universiteit, Amsterdam
- **Radhika Apaiah***, Product Design and Quality Management Group, Wageningen University
- **Fred Beekmans**, Agrotechnology & Food Innovations, Wageningen University and Research Centre
- **George Beers**, Agricultural Economics Research Institute, The Hague
- **Tiny van Boekel***, Product Design and Food Quality Management Group, Wageningen University
- **Joop de Boer***, Institute for Environmental Studies, Vrije Universiteit, Amsterdam
- **Jos Cornelese**, Ministry of Agriculture, Nature and Food Quality, The Hague
- **Paul Diederen**, Agricultural Economics Research Institute, The Hague
- **Hanneke Elzerman***, Product Design and Quality Management Group, Wageningen University
- **Cees de Graaf**, Division of Human Nutrition, Wageningen University
- **Dolf de Groot***, Environmental Systems Analysis Group, Wageningen University
- **Harry Gruppen**, Laboratory of Food Chemistry, Wageningen University
- **Randolph Happe**, TNO Nutrition and Food Research, Zeist
- **Martine Helms***, Institute for Environmental Studies, Vrije Universiteit, Amsterdam
- **Eligius Hendrix**, Operations Research and Logistics, Wageningen University
- **Lynn Heng***, Laboratory of Food Chemistry, Wageningen University
- **Claudia Herok***, Agricultural Economic Research Institute, The Hague
- **Annet Hoek***, Division of Human Nutrition, Wageningen University
- **Rob Hoppe**, Policy Studies, Twente University, Enschede
- **Ekko van Ierland***, Environmental Economics Group, Wageningen University
- **Leo Jansen**, Platform Sustainable Development, Delft University of Technology
- **Wim Jongen**, Product Design and Quality Management Group, Wageningen University

- **Lia Kemper**, Technology Foundation STW, Utrecht
- **Michiel Keyzer**, Centre for World Food Studies, Vrije Universiteit, Amsterdam
- **Pieter-Jan Klok**, Policy Studies, Twente University, Enschede
- **Gerrit van Koningsveld**, Laboratory of Food Chemistry, Wageningen University
- **Onno Kuik***, Institute for Environmental Studies, Vrije Universiteit, Amsterdam
- **Egbert Lantinga***, Biological Farming Systems Group, Wageningen University
- **Anita Linnemann***, Product Design and Quality Management Group, Wageningen University
- **Hans Linsen**, Programme Committee Chairman
- **Pieternel Luning**, Product Design and Quality Management Group, Wageningen University
- **Gerrit Meerdink**, Product Design and Quality Management Group, Wageningen University
- **Max Merbis**, Centre for World Food Studies, Vrije Universiteit, Amsterdam
- **Chris Mombers**, Technology Foundation STW, Utrecht
- **David Niemeijer***, Environmental Systems Analysis Group, Wageningen University
- **Francesca O'Kane***, Product Design and Quality Management Group, Wageningen University
- **Joop van der Pligt**, Department of Psychology, University of Amsterdam
- **Krit Raemakers***, Laboratory of Plant Breeding, Wageningen University
- **Jacques Roozen**, Laboratory of Food Chemistry, Wageningen University
- **Dick Stegeman***, Agrotechnology & Food Innovations, Wageningen University and Research Centre
- **Frank van Tongeren**, Agricultural Economic Research Institute, The Hague
- **Manolis Tzitzikas***, Laboratory of Plant Breeding, Wageningen University
- **Pier Vellinga**, Institute for Environmental Studies, Vrije Universiteit, Amsterdam
- **Wim Verbeke**, Department of Agricultural Economics, Ghent University, Belgium
- **Harmen Verbruggen**, Institute for Environmental Studies, Vrije Universiteit, Amsterdam

- **Johan Vereijken***, Agrotechnology & Food Innovations, Wageningen University and Research Centre
- **Marike Vijver***, Policy Studies, Twente University, Enschede
- **Jean-Paul Vincken***, Laboratory of Plant Breeding, Wageningen University
- **Richard Visser***, Laboratory of Plant Breeding, Wageningen University
- **Jan Vos***, Crop and Weed Ecology Group, Wageningen University
- **Henk Waaijers**, Division of the Social Sciences, Netherlands Organisation for Scientific Research, The Hague
- **Lia van Wesenbeeck***, Centre for World Food Studies, Vrije Universiteit, Amsterdam
- **Justus Wesseler**, Environmental Economics Group, Wageningen University
- **Frank Willemsen***, Institute for Environmental Studies, Vrije Universiteit, Amsterdam
- **Xinyou Yin***, Crop and Weed Ecology Group, Wageningen University
- **Xueqin Zhu***, Environmental Economics Group, Wageningen University

PROFETAS SPONSORS

- **Netherlands Organisation for Scientific Research** (NWO)
- **Technology Foundation STW** (which is financed by NWO and the Netherlands Ministry of Economic Affairs)
- **Netherlands Ministry of Agriculture, Nature and Food Quality** (LNV)
- **Boekos BV**
- **Cebeco Group UA**
- **DSM Food Specialties**
- **Quest International Nederland BV**
- **Unilever Research Vlaardingen**
- **Wageningen Centre for Food Sciences**

FURTHER INFORMATION: **WWW.PROFETAS.NL**

LIST OF ABBREVIATIONS

AGE	Applied General Equilibrium
BCT	Bud-Containing Tissue
BSE	Bovine Spongiform Encephalopathy
CAP	Common Agricultural Policy of the European Union
DPSIR	Driving force – Pressure – State – Impact – Response
DSR	Driving force – State – Response
EU	European Union
FAO	Food and Agriculture Organization of the United Nations
FIC	Food Insecure Countries
FNC	Food Neutral Countries
GATT	General Agreement on Tariffs and Trade
GDP	Gross Domestic Product
GIS	Geographic Information System
GM	Genetically Modified
GMOs	Genetically Modified Organisms
HPLC	High Performance Liquid Chromatography
J	Joule
LCA	Life Cycle Analysis
MS	Mass Spectrometry
NGO	Non-Governmental Organization
NL	Netherlands
NPFs	Novel Protein Foods
NSPs	Non Starch Polysaccharides
NWO	Netherlands Organisation for Scientific Research
OECD	Organisation for Economic Cooperation and Development
PAGE	Poly Acrylamide Gel Electrophoresis
pH	Acidity
PROFETAS	Protein Foods, Environment, Technology And Society
PSR	Pressure – State – Response
R&D	Research & Development
RNA	Ribonucleic acid
SARS	Severe Acute Respiratory Syndrome
STD	Sustainable Technology Development
TRIPs	Trade-Related Intellectual Property Rights
TVP	Texturised Vegetable Protein
VAT	Value Added Tax
WCED	World Commission on Environment and Development
WTO	World Trade Organization

CHEMICALS AND CHEMICAL SUBSTANCES

CFC	Chlorofluorocarbon
CO_2	Carbon dioxide
CH_4	Methane
N	Nitrogen
NEM	N-ethylmaleimide
NH_3	Ammonia
NH_4^+	Ammonium
N_2O	Nitrous oxide
NO_x	Nitrogen oxide(s)
P	Phosphorus
P_2O_5	Phosphorus oxide
SH	Sulphydryl
SO_2	Sulphur dioxide
SDS	Sodium dodecyl sulphate

CHAPTER 1

BACKGROUND, AIMS AND SCOPE

Harry Aiking & Joop de Boer

1.1 FOOD, AGRICULTURE AND SUSTAINABILITY

These are unique times. Never before have the world population and the standard of living been so high and, in fact, both are still on the rise. Consequently, human activities are at unprecedented levels with regard to volume, rate of change and consequences, and among such activities food production takes a unique place (Vellinga and Herb, 1999). This book deals with stepwise changes – called societal *transitions* – to make food production and consumption more sustainable. That is of utmost importance, because the current ways of food production and consumption cause far too much environmental pressure to ensure a sustainable food supply in the future. Moreover, this state of affairs is more likely to rapidly deteriorate than to improve by itself, because the very ways in which the food system is organized may inhibit the very changes that could make it more sustainable (Tansey and Worsley, 1995; Millstone and Lang, 2003). Although there are no easy ways to stimulate the necessary transition and actual "transition management" seems remote, there is occasionally a "window of opportunity" for a decisive improvement. If the economic and political conditions might open this window, the right options should be available at the right time. In order to be prepared for a window of opportunity to open, it is of vital importance that decision makers from governments and industries be equipped with one or more sustainable alternatives, which are both feasible from a technological point of view and acceptable in terms of the main values in society.

The general aim of the PROFETAS (PROtein Foods, Environment, Technology and Society) research programme – the subject of this book – is to ease the window open by developing and evaluating potential options for a transition towards sustainability. More in detail, the programme's aim is to assess whether a transition in protein production and consumption – shifting from meat to plant proteins – may provide the necessary alternative. In the programme, about two dozen multidisciplinary researchers have studied the environmental, technological and societal feasibility of the transition from the currently predominant consumption of animal products towards a society in which protein-rich products based on plant proteins – called Novel Protein Foods (NPFs) – have replaced a significant part of the meat protein products. The projects are concerned with the environmental sustainability of protein-rich foods and the comparative evaluation of meat with NPFs (Chapter 2), the technological feasibility of making a tasty product in view of the crop, the proteins and the production chain (Chapter 3), the acceptance of NPFs by consumers (Chapter 4), and the transition from national and international economy and policy perspectives (Chapter 5). The present book puts the results in a broader perspective. From the original position and background of the programme (Chapter 1), finally, it identifies and analyses emerging options (Chapter 6) and summarizes the insights and tools delivered (Chapter 7).

Global environmental pressure

The relevance of developing alternatives to the current ways of food supply is evident since, on the one hand, food is the most basic of commodities and, on the other hand, a major proportion of global environmental pressure is caused by food-related human activities (Alexandratos, 1995; Brown, 1996). The awareness of such a tight coupling between food, agriculture and sustainability dates back to the 1980s, when sustainable development became an overarching policy objective for all nations. After 20 years of debate on how to manage the earth's resources, it was a milestone that the World Commission on Environment and Development (WCED), also known as the Brundtland commission (Brundtland, 1987), was able to establish clear linkages between (1) global environmental deterioration, (2) poverty, and (3) rapid population growth. In view of this extremely negative prospect, the WCED stated that "(h)umanity has the ability to make development sustainable – to ensure that it meets the needs of the present without compromising the ability of future generations to meet their own needs" (Brundtland, 1987).

The WCED statement is often used as a definition of sustainable development. It links the environment's ability to meet present and future needs with theories of social justice – both within and between generations – as a basis for ecological, economic and social aspects of sustainability (Langhelle, 2000). Given its high level of abstraction and generality, however, the concept should be used with caution. For example, it is often not possible to fully specify what "sustainable development" ideally means at the level of a particular product, because the concept refers to long-term checks and balances between ecological, economic and social processes at the level of society as a whole (Pezzey, 1992). In addition, there may be an abysmal gap between producers' and consumers' understanding of sustainability. Kloppenburg et al. (2000) note that several large companies, such as DuPont and Monsanto, describe sustainability in terms of "ecologically sound, economically viable and socially acceptable." In contrast, consumers may use "sustainability" as a kind of shorthand for the "green and good" to indicate production systems associated with a broader range of attributes, such as community-based efforts to build healthy, just and local food systems.

Despite such difficulties, there are still many attempts to improve the substantiation of sustainability claims at the level of production methods and products. Instead of claiming very general differences in sustainability, it appears more feasible and attractive to work up from a low level of abstraction with relatively concrete topics, identifying significant "ills" to be escaped (e.g. dependence on pesticide use) or specific "ideals" to be approached (e.g. biochemical efficiency). As the political scientist Lindblom (1990) notes, people who are confronted with social problems can opt to eliminate some of the most severe forms of poverty, although they may not have a broadly shared opinion on the ideal distribution of income and wealth. In other words, it is often easier for a heterogeneous society to agree on the ills to be avoided (e.g. poverty) than on the ideals to be achieved (e.g. the ideal income distribution). Nonetheless, the pursuit of sustainability may involve a combination of avoidance and approach. It should not be understood as a requirement to maintain a static situation, but as a challenge to preserve the resilience and adaptability of the natural systems that form the basis of social and economic development.

Can technology and society vent the environmental pressure?

The relationships between food production, environment and society are complex. The evolution of agriculture has both shaped and been shaped by world population growth (Evans, 1998). Presently, the ecological basis of food production in many regions, including the Netherlands, is threatened by chains of food-related human activities: crops are grown, processed, turned into food products and transported in ever-larger volumes, with ever-increasing impacts on the environment (Hoffmann, 2001; Millstone and Lang, 2003; Tilman et al., 2002). The main problems are inappropriate pest control, reduction in biodiversity, soil erosion, as well as inefficient nutrient, water and energy use. Currently, over one-third of all ice-free land area is used for food production, along with over three quarters of the available freshwater (Smil, 2002: 239). Two important driving forces are the growth of the world population, which will continue for several more decades, and the increasing consumption of meat proteins in many parts of the world. While the world population doubled during the second half of the 20th century, its appetite for meat quadrupled, requiring over 40% of the world grain harvest to be fed to livestock (Evans, 1998).

The production of meat is increasingly showing drawbacks in several ways. During recent meat crises, the meat production system has proved increasingly prone to animal diseases such as swine fever, BSE, foot-and-mouth and avian influenza, strongly affecting both animal welfare and consumer perception. With increasing frequency, animal disease outbreaks hit global meat exports and increasingly cause world trade losses through export bans or market constraints, in addition to costs of public disease control measures and losses to producers and consumers through destabilized markets and fluctuating prices. On 2 March 2004, the UN Food and Agriculture Organization (FAO) estimated about one-third of global meat exports, or 6 million tons, to be affected by animal disease outbreaks. With the value of global meat and live animal trade estimated at $33 billion (excluding EU intra-trade), this could amount to world trade losses of up to $10 billion throughout 2004. Apart from economic damages, in some cases (Campylobacter, BSE, avian influenza), human health is affected, as well.

Even more importantly, however, meat production is particularly environment-unfriendly, which is primarily due to the inherently inefficient conversion of plant protein to meat protein. Direct human consumption of the plant proteins requires only a fraction of the input of natural resources since, although this "conversion factor" depends on the type of animal and the production conditions, on average, 6 kg of plant protein is required to yield 1 kg of animal protein (Pimentel and Pimentel, 2003; Smil, 2000). Consequently, meat production is responsible for a disproportionate share of environmental pressure: resource utilisation (such as land area, biodiversity, freshwater), as well as pollution (eutrophication, pesticides, climate change). When striving for sustainable ways of food production and consumption, therefore, the protein chain is an excellent starting point (Bradford, 1999; Delgado *et al.*, 1999; Gilland, 2002; Grigg, 1995b; Millstone and Lang, 2003; Smil, 2000; Weaver *et al.*, 2000). A promising solution addressing environment and human health as well as animal welfare issues may be created by a reduction of meat production and replacement of the meat proteins with plant protein products. Several economic arguments (Seidl, 2000; White, 2000) indicate, however, that actual practice may not be as straightforward as theory suggests.

One of the gaps between theory and practice is the technological innovation necessary to create marketable plant protein products. Alternatives to meat protein based on plant proteins have, in fact, been around for thousands of years. Some traditional Asian foods such as tofu and tempeh belong in this category. Today, however, these traditional, as well as more recently developed non-meat protein foods, are poorly represented within the diet of most people, and they are widely perceived as inferior in taste and texture to foods based on meat proteins. Nevertheless, new opportunities exist to develop a wide range of products based on proteins from plants, such as peas, beans and lucerne, or from micro-organisms, such as *Spirulina* or *Fusarium* species (Weaver et al., 2000). Since it is felt that such new products should stand as protein sources in their own right, rather than be compared to meat (as in "meat substitutes"), they are generically called Novel Protein Foods (NPFs).

Developing NPFs may require many innovations and these will not take place automatically as the outcome of a linear process. Generally, the search for, development of, and adoption of new production methods and products are the outcome of interactions between (a) capabilities and incentives generated within industries and (b) broader causes external to individual industries. According to the economist Dosi (1988), the latter category includes factors such as the state of science in different branches, facilities for the communication of knowledge, conditions controlling consumer promptness or resistance to change, market conditions, macroeconomic trends, especially in their effects on changes in the relative prices of inputs and outputs, as well as public policies (e.g. tax codes, patent laws, industrial policies).

Convergence of problems, solutions, and political events

Whether there will be opportunities to make food production more sustainable depends not only on the strength of the need to develop alternatives but also on timely reminders of their very existence. The notion of a "window of opportunity" is based on the observation that decisions of businesses and governments are often influenced by both the *content* and the *timing* of issues (Cohen et al., 1972; Kingdon, 1984; Nill, 2002). Decision makers in economic and political organizations often have to deal with various streams of events, namely (1) streams of problem-related events, such as news about food scares or environmental degradation, (2) streams of solution-related events, such as reports on research and development, and (3) streams of management-related decision opportunities, such as decisions on the annual R&D budgets. These streams are to a certain degree independent of each other, because they flow along their own schedules and according to their own rules. On occasion the streams converge, however, and then opportunities may present themselves to link one or more problems to one or more solutions.

The convergence of problems, solutions and political events may create a window of opportunity for a transition from meat to novel protein foods (see

Figure 1-1). Whether and how the window will be used depends on the technical feasibility of the transition and the degree to which it is compatible with the values that prevail in the most directly involved economic and political organizations. Specialists in the relevant policy areas, such as researchers and members of staff, may play a key role in these organizations. As the political scientist Kingdon (1984) notes, specialists often nourish a shortlist of solutions and those solutions that do not agree with their values have less chance of survival than those that do. What the specialists need if the window opens is a viable alternative that they can offer to policymakers. If an alternative has not been worked out yet, its chances of success will decrease.

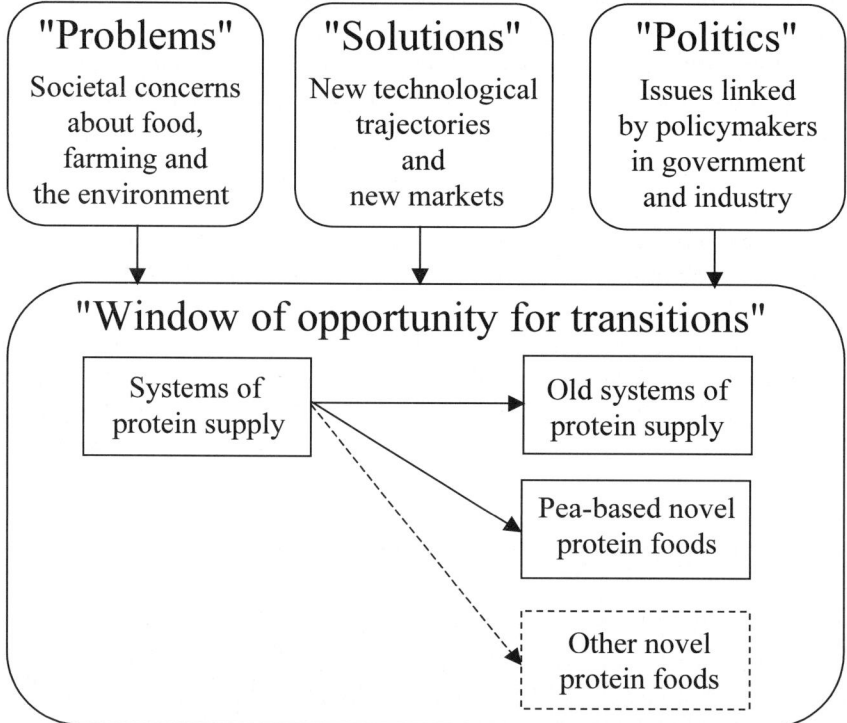

Figure 1-1. The convergence of problems, solutions, and political events may create a window of opportunity for a transition to novel protein foods.

Figure 1-1 illustrates the relevance of distinguishing between factors that contribute to the opening of a policy window and factors that improve the chances that the open window will be used in a particular way, for example, towards a transition (towards sustainability) in protein supply. The latter refers especially to the type of competition between old and new technologies. Novel protein foods will be the product of a new technology, which has to compete with the old technology of meat-based protein supply. Under these circumstances, the established technology may seriously hamper the development of promising innovative products. In addition, there may be a competition between several kinds of NPFs, such as pea-based versus grass-based protein foods. In such a competition, first movers on the market may gain a decisive advantage but they will also run significant risks. Accordingly, the market system will not simply guarantee that the most sustainable technology wins.

In short, relieving the environmental pressure of food production for a growing world population – which might approach 9 billion by 2050 – is a challenge for all countries. To make food production and consumption more sustainable, a stepwise improvement is required, coined a "transition" (Weaver *et al*., 2000). In the past, many transitions have occurred, but they always evolved passively, as products of a multitude of chance factors. It is currently thought in the Netherlands that active transition "management" should be sought by the government (Van Wijk and Rood, 2002). However, many actors are involved, all of which will perceive their own barriers and opportunities. For new consumer products, such as NPFs, much may be learned from the butter to margarine transition (Grigg, 1995b). Like the protein transition, it also involved replacing an animal product with a plant product. Due to dairy industry (and – in their wake – governmental) obstruction its breakthrough was delayed for about one century. Without the boost provided by the intervening Second World War margarine might still be a marginal product today (Van Stuijvenberg, 1969).

1.2 THE PROFETAS APPROACH

In 1989, the Dutch government embraced its responsibility towards sustainability in its First National Environmental Policy Plan. One initiative under the plan was to establish an interministerial programme elucidating and facilitating the role of technological innovation towards sustainability: the programme for Sustainable Technology Development (STD) (Weaver *et al*., 2000). This programme used the technique of backwards reasoning (so-called "backcasting") to identify sustainable alternatives for important societal needs. After first creating a future vision with significant challenges for environmental improvements, experts looked back on how this desirable future could be achieved, and what technologies should be developed in consequence. Generally, the aim was a 20-fold improvement of the environmental performance of the system. Several case studies were examined, some of which were in the domain of nutrition, including one on NPFs. Two of the conclusions of this desk study were that by 2035 (a) 40% of meat consumption should be replaced to achieve the desired 20-fold reduction of environmental impact and that (b) NPFs constituted a feasible option, not by imitating whole cuts of meat (such as chops or steaks), but as ingredients (such as constituents of sauces, soups, sausages and snacks).

Against this background, the PROFETAS research programme aims to assess from a more practical perspective and more in detail whether and how the transition of protein production and consumption may contribute to a more sustainable system of protein supply. This requires a comprehensive approach that addresses issues such as technological innovation, consumer demands, and long-term environmental changes. The programme builds on two conclusions drawn from the desk study mentioned above (Weaver *et al.*, 2000). First, predicting actual products 10-40 years in advance is not feasible and, therefore, it is better to develop now the methodology and tools to facilitate problem solving in the future than to develop solutions for presently perceived future problems. Second, trying to mimic meat chops (such as steaks) with plant proteins is not feasible. It is hard to turn the often *globular* plant storage proteins into products mimicking the highly ordered structure of *fibrous* muscle proteins making up the bulk of meat. In the Netherlands, in the late 1960s, in fact, it was tried in vain to market mock steaks, which contained texturised vegetable protein (TVP) derived from soy (Baudet, 1986). On the international market TVP is still available, but as a meat substitute for humans it constitutes a marginal product. Therefore, NPFs serving as protein-containing diet ingredients should be developed in a consumer-directed way.

Building on both the "toolkit" and the "ingredient" philosophy described above, PROFETAS examines the entire protein chain (from primary production via processing and consumption to waste), rather than concentrating on the primary production, as has been the focus of most environmental research in this area (such as life cycle analysis of food products). It gives a predominant role to consumer preferences when designing and evaluating alternative protein chains. And, last but not least, it develops a multidisciplinary (political, social, economic, technological, environmental, ecological, and chemical) approach to the design and evaluation of alternative protein production options and their environmental, economic and social impacts.

The PROFETAS projects compare the NPF sector with the intensive livestock industry in the Netherlands and the EU (see Figure 1-2). Because sustainability is a global objective, environmental, economic and social issues are studied in the context of global development. The common object

of study is the triple hypothesis that a shift in the Western diet from meat protein to plant proteins is:
1. environmentally more sustainable than present trends,
2. technologically feasible,
3. socially desirable.

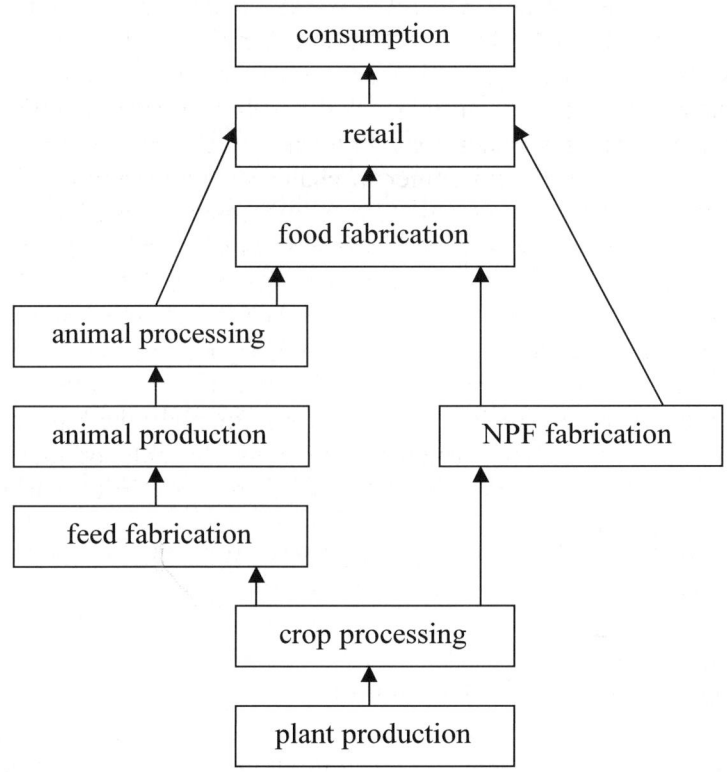

Figure 1-2. The protein supply chains (meat vs. Novel Protein Foods) to be compared.

The hypotheses are studied in a set of 15 projects in different scientific disciplines. Firmly based on disciplinary and multidisciplinary research, the projects are well equipped to outline problems, indicate options, state preferences from a scientific point of view and develop tools.

The relations between the three hypotheses provide the overarching PROFETAS structure (see Figure 1-3). Environmental sustainability and technological feasibility provide prerequisite partial input for social desirability. Subsequently, the latter is translated into policy options for policymakers from government and industry. They can use these tools in order to develop, implement and evaluate policies. Thus, two layers of aggregation condense multidisciplinary information in a funnel-shaped approach.

In order to focus and synchronise the programme, two reference production and consumption chains were devised (Figure 1-2). The pig chain was selected for the common reference meat chain mainly because of the absence of secondary products such as milk or eggs. For the plant protein chain the pea (*Pisum sativum L.*) was selected because of the protein content, the ability to grow in Western Europe, the absence of unwanted substances in the pea and the available expertise, etc. (see Section 6.2 for details).

The basic structure of the programme was implemented initially by defining five PhD student research projects in the area of environmental sustainability and five in technological feasibility, plus five Postdoc research projects in the area of social desirability. In actual practice, some modifications to the basic structure were made. The final translation of the results into options for policymakers from government and industry, respectively, became the responsibility of two programme coordinators.

PROFETAS is aimed at assisting policymakers from (a) government, and (b) industry with tools and arguments, as indicated above, providing them with "options for policy". However, life is not as simple as that. It has to be taken into consideration that, without exception, policies have side effects. For example, policies in seemingly unrelated areas – say, employment policy – are likely to have secondary effects on sustainability, and on a potential protein transition, for that matter. By the same token, agricultural policy can have unanticipated effects. Thus, after World War II the main goal of agricultural policy in the EU was food security. This resulted in subsidised surpluses, sustained Third World poverty, and continually increasing environmental impacts of food production.

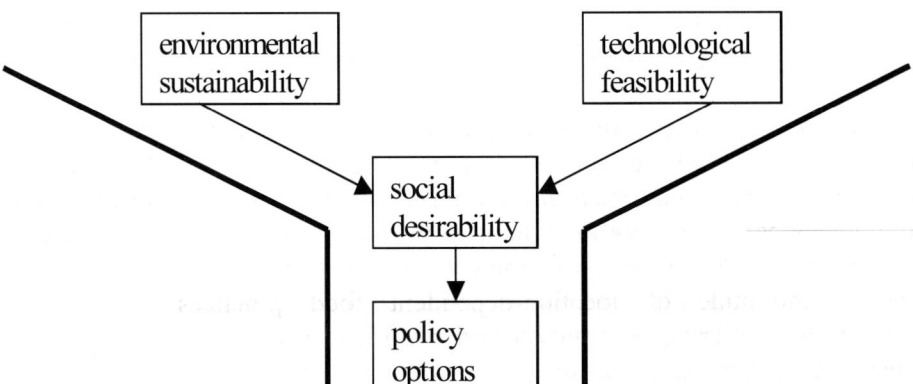

Figure 1-3. The funnel approach towards multidisciplinary aggregation.

Presentation of the results

Since all constituent projects have specific goals and contribute to common goals, there is no self-evident or linear format for reporting the PROFETAS results. Therefore, the present report follows the classification of projects according to Figure 1-3. This results in chapters on environmental sustainability (Chapter 2), technological feasibility (Chapter 3) and social desirability (Chapters 4-5). Each project describes its specific goal, outlines its relevant issues, and reports the results and the gaps remaining. At the end of each chapter, the implications for the most relevant of the three PROFETAS hypotheses are discussed. Subsequently, interesting emerging options are described and analysed in Chapter 6. The concluding Chapter 7 summarises the insights and tools delivered, and further insights and tools required for a transition. Furthermore, it provides an assessment of transition feasibility, implications for stakeholders and options for policy.

1.3 ORIGINS OF DIET PROTEINS IN EUROPE (EU-15)

For a research programme such as PROFETAS it is useful to develop some insight into the background of diet proteins and their origins (plant, dairy, fish or meat), within the boundaries determined by nutrition and health. The aim will be to survey the European (EU-15) protein consumption in its full variety, while trying to explain national differences in the quantity and source of protein. This will give some insight into (a) what differences exist between national diets, so what range of sustainability gains can potentially be expected and (b) what societal differences may have to be taken into account in the pursuit of food sustainability.

Diet differences may stem from ecological, economic and/or cultural factors (Grigg, 1995a; Grigg, 1995b). Some clues on current developments in the world can be derived from the transformation of the European diet in the 20th century. In their historical analysis, Teuteberg and Flandrin (1999) indicate that the proportion of animal proteins and fats has increased dramatically in all countries – except forerunner France – after World War II, possibly as part of a more general homogenisation of food consumption patterns across Europe. Such a homogenisation may significantly reduce the age-old multitude of location-dependent food practices shaped by differences in climate, vegetation and historical development (Montanari, 1994). Traditionally, the relative proportion of different meats in the diet fluctuated with time and varied considerably from country to country, and the same applied to fish. Differences in the local retail cost per gram of protein and differences in income often result in a positive spatial correlation between, on the one hand, local income and, on the other hand, the more expensive types of protein such as meat (Grigg, 1995b). Relatively recent trends of increases in agricultural productivity, decreasing differences in national income, increased trade in food and an increasing globalisation of eating habits have greatly diminished national diet variation. However, there still appear to be wide disparities between the consumption of certain food items within the countries of the Single Market (Askegaard and Madsen, 1998; European Communities, 2001).

Although it is far beyond the scope of this book to explain all the protein-related differences between countries, some sociological observations should be mentioned. Since food choices are no longer tied to the "natural rhythm" of an agricultural society, they have diverged into a socio-economic field of their own (Ilmonen, 1991). Although socio-economic development can bring pervasive cultural changes, it has been observed that cultural values, such as embedded in the cultures of predominantly Protestant and Catholic countries, are an enduring influence on society and may allow a development that is path-dependent (Inglehart and Baker, 2000). In this case, path dependency means that socio-economic development has been different in countries from the Protestant and from the Roman Catholic cultural zone in Europe. To avoid any misunderstandings, this observation refers to processes at the level of countries; it does not necessarily mean that Protestants and Catholics from the same country will differ in their diets. In the case of food, internationalisation seems to take place primarily on the level of ingredients in the "planetary supermarket", leaving much room for already existing categorisations and rules on what can be eaten with what and when (Askegaard and Madsen, 1998; Fischler, 1999). It is, in fact, the level of ingredients that is of primary importance from the perspective of sustainability.

For a protein consumption inventory of EU-15, two sources of data are valuable. First, FAOSTAT supply data will be employed to compare supply patterns as indicators of consumption patterns. The data refer to national per capita supply at the retail level (supply = production + imports - exports). The second dataset is the compilation of national household budget surveys, provided by Eurostat (2005). Current FAOSTAT data cover the period from 1961 to 2002, but we will mainly use data on 1999, which is the most recent year of the Eurostat household budget survey. As FAOSTAT used to combine the data on Belgium and Luxembourg, the actual number of distinguished "countries" is 14, rather than 15.

Table 1-1 specifies the main sources of daily protein *supply* (g/person) in 14 European countries. Compared to the overall mean of 108.6 g dietary proteins, it appears that, on average, plant-derived proteins provide the smallest part, ranging from 32.6 g (The Netherlands) to 53.3 g (Greece). Of all plant sources, cereals (mostly wheat) are the largest suppliers of plant protein in all EU countries. Vegetables and potatoes are the second and third most important suppliers of plant protein, but their contributions are relatively small. Interestingly, the present contribution of pulses is surprisingly low in all countries, considering their age-old importance as a protein-delivering foodstuff until 1945 (Smil, 2000).

Table 1-1. Main sources of daily protein supply (g/person) in 14 European countries in 1999 (* including the small contributions of "other sources"). Source: (FAO, 2004).

Source	Mean	Min	Country	Max	Country
Cereals	26.2	17.6	The Netherlands	35.5	Italy
Potatoes	3.4	1.6	Italy	5.5	Portugal
Pulses	1.6	<0.1	4 countries	3.7	Spain
Tree nuts	0.2	<0.1	12 countries	1.7	Greece
Oil crops	0.4	<0.1	10 countries	1.6	Germany
Vegetables	4.1	2.1	Finland	7.7	Greece
Fruit	1.4	1.0	Finland	2.5	Greece
Stimulants (incl. coffee)	1.4	<0.1	Greece	2.6	Denmark
Alcoholic beverages	0.9	<0.1	6 countries	2.6	Ireland
Beef & veal	7.6	4.6	Germany	10.4	Italy
Mutton & goat	1.3	0.1	Finland	5.4	Greece
Pork	12.5	6.7	United Kingdom	22.7	Austria
Poultry	7.9	4.6	Sweden	11.7	Ireland
Offal	2.4	0.5	Denmark	9.6	Ireland
Milk (incl. cheese)	22.4	13.9	Spain	28.2	The Netherlands
Eggs	3.7	2.1	Ireland	5.0	France
Fish/Seafood	7.2	2.9	Austria	15.7	Portugal
Total* plant protein	41.9	32.6	The Netherlands	53.3	Greece
Total* animal protein	66.7	55.3	United Kingdom	76.2	France
Total* protein	108.6	95.8	Germany	118.9	Portugal

After summation of the five meat categories, this source is generally the largest supplier of animal proteins, but its average diet contribution (31.7 g) is not much higher than that of cereals (26.2 g). Different types of meat are not preferred equally in all countries. Pork, in particular, though the largest meat protein supplier in general, shows large variations in its relative contribution. Directly behind meat, milk (including cheese, but excluding butter) is the second most important supplier of animal proteins in most EU countries. The fish/seafood food group is on average relatively unimportant as a source of protein, although it is responsible for a large variation between countries.

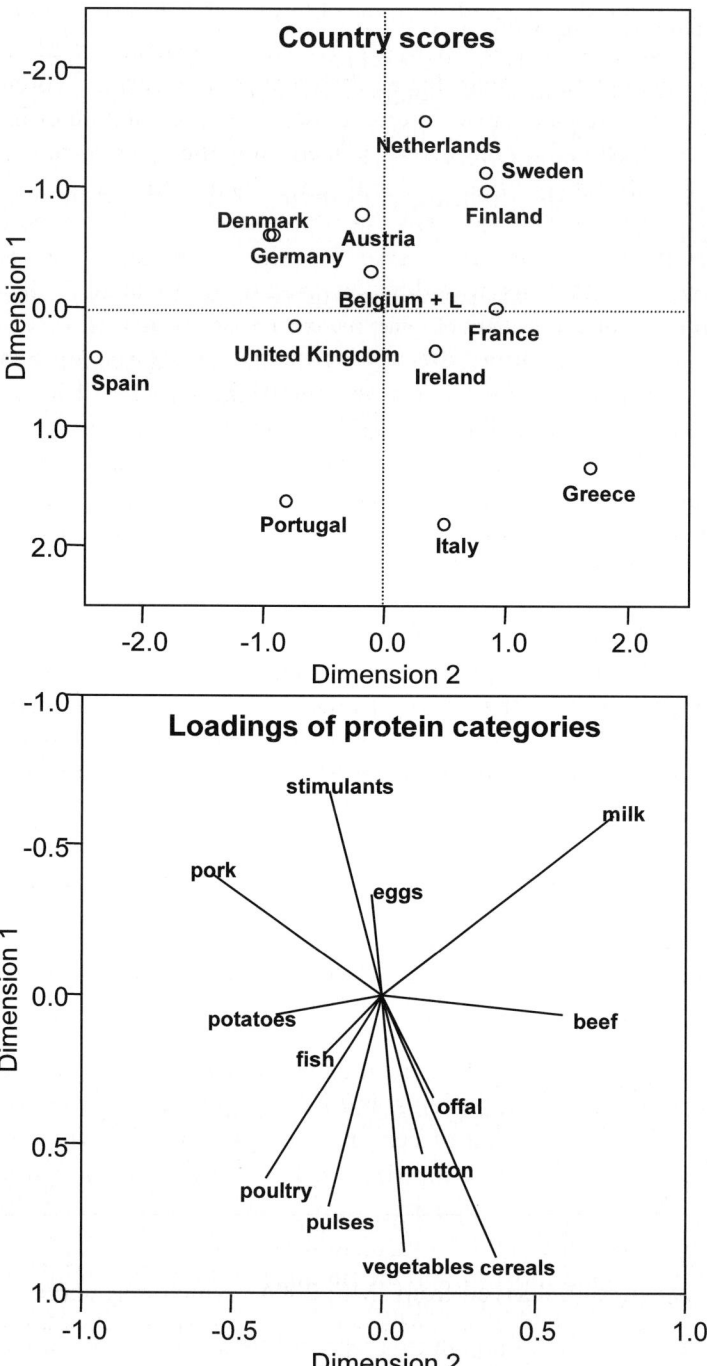

Figure 1-4. Plots of the statistical analysis: country scores (top) and relationships between the protein categories (bottom) after rotation and mirroring of the original output.

A more systematic way of dealing with the range in contributions of the various protein sources is to apply statistical analysis. This technique condenses the various protein-based differences between the countries in a two-dimensional space. Those countries that resemble each other in terms of their main protein sources are placed together; countries that are significantly different are placed at a distance from each other.

The two dimensions are based on the statistical relation between the protein sources, which may show positive, negative or negligible correlations. The analysis was done by means of the statistical procedure CAPTA for performing optimal scaling; only those protein sources of Table 1-1 were included that contribute on average at least 1 g protein with a range of at least 2 g between the lowest and the highest levels. The top part of Figure 1-4 plots the positions of the 14 countries on the two dimensions. Interestingly, the statistical output largely corresponds with a map of Europe. Because most people are used to a map in which the Mediterranean countries are arranged on a horizontal line from left to right, Figure 1-4 shows the output after rotation and mirroring, which do not influence the relationships between the depicted elements.

The vertical dimension of Figure 1-4 separates countries with high supplies of protein derived from, in particular, milk from countries with high supplies of protein provided by vegetables and cereals. The main contrast is that between the Netherlands and Sweden, on the one hand, and Portugal, Italy and Greece, on the other hand. The horizontal dimension separates countries with a relatively high share of beef and milk protein from countries that depend on other protein sources, such as poultry. The exceptional position of Spain is largely determined by the relatively low supply of milk protein in that country. Interestingly, the bottom part of Figure 1-4 shows that the loadings of the main varieties of meat (e.g. pork, beef and poultry) point in different directions, indicating that each country may have its own preference for one of these three protein sources without neglecting the others.

Further analysis demonstrates that the vertical dimension of Figure 1-4 is crucial in separating the 14 countries in relation with a number of ecological, economic and/or cultural factors. These are (1) the countries' average latitude (derived from CIA, 2004), which may serve as a proxy for differences in ecological (climate, vegetation, fauna) and historical development, (2) their Gross Domestic Product per capita (current prices, US dollar per person, derived from IMF, 2004), which may serve as a proxy for differences in income and corresponding economic development, (3) their meat expenses per household member (i.e. Eurostat (2005) meat purchasing data), and (4) their proportion of Protestant inhabitants (derived from the European Values Survey (e.g. Bréchon, 2004)). The latter refers to the boundaries between – what marketers (Askegaard and Madsen, 1998) call – a more ascetic, Protestant food culture and a more indulgent Catholic pattern. The following analyses may clarify these relationships.

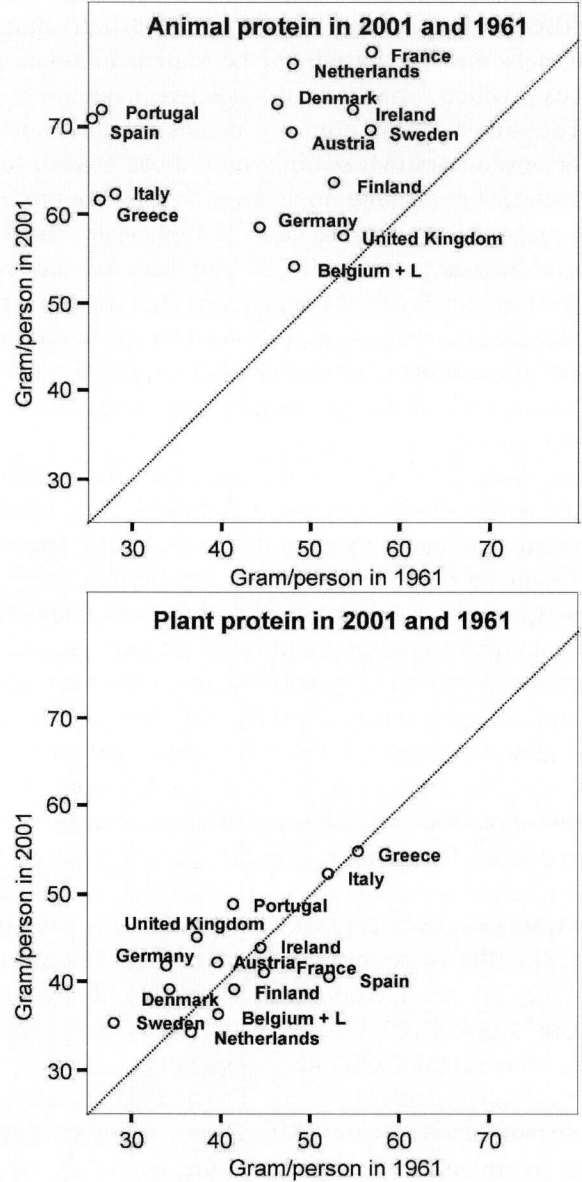

Figure 1-5. National averages of daily animal protein supply and plant protein supply in the years 2001 and 1961.

A methodologically proper way to clarify the combined effects of the countries' average latitude and their national income is to put the protein supply data in the perspective of rising incomes since the 1960s. From the FAOSTAT supply data it is clear that the average total protein supply in the combined EU-15 member states has increased gradually and consistently during 1961-2001, and that this increase was attributable to a rise in animal proteins exclusively. Figure 1-5 summarizes this development by displaying national averages of animal protein supply and plant protein supply in the years 2001 and 1961.

The figure clearly shows that the supply of *plant* proteins has been remarkably stable over the past forty years, though the supply of *animal* proteins has increased significantly. As a result, in all countries the present supply of animal proteins is larger than the supply of plant proteins. The supply of plant proteins shows a decreasing trend from southern to northern countries, once linked to the phenomenon of a high-plant and low-meat, "Mediterranean" type of diet. However, especially the countries that once showed a "Mediterranean" type of diet (i.e. Greece and Italy) have increased their supply of animal protein in the recent past.

As the rising incomes did not affect the North-South trends in plant protein supply, historical differences in availability and culture may still be an important cause underlying present differences in demand. In the case of meat protein, rising incomes and all its associated economic changes may almost have completed the diet transformation that started after World War II. As Grigg (1995a) notes, by the 1960s most countries in western Europe – save the Mediterranean area – had already gone through the transformation. Given the S-shaped relationship between national income and meat consumption and the position of all EU-15 countries in the saturated part of the curve (GDP > $ 10,000) (Keyzer *et al.*, 2005), it is to be expected that the consumption of meat will not increase much when incomes rise further.

However, the amount of money consumers spend on food seems partly dependent on cultural factors. The different food cultures of predominantly Protestant and Catholic countries may allow a path-dependent development. Figure 1-6 shows that increasing annual expenditures of households are related to increasing expenditures on meat, but that the average meat expenditures are significantly higher among countries with a Catholic background than among the Protestant countries (country names underlined). In fact, Figure 1-6 clearly demonstrates that even those Protestant countries with a high level of overall expenditures did not spend as much on meat as some Catholic countries with a much lower level of overall expenditures. Notably, these differences should be understood at the level of countries and cultural zones. Sociological research has shown that the basic values of Catholics in mixed countries such as Germany and the Netherlands resemble those of their Protestant countrymen more than they resemble Catholics in other countries (Inglehart and Baker, 2000).

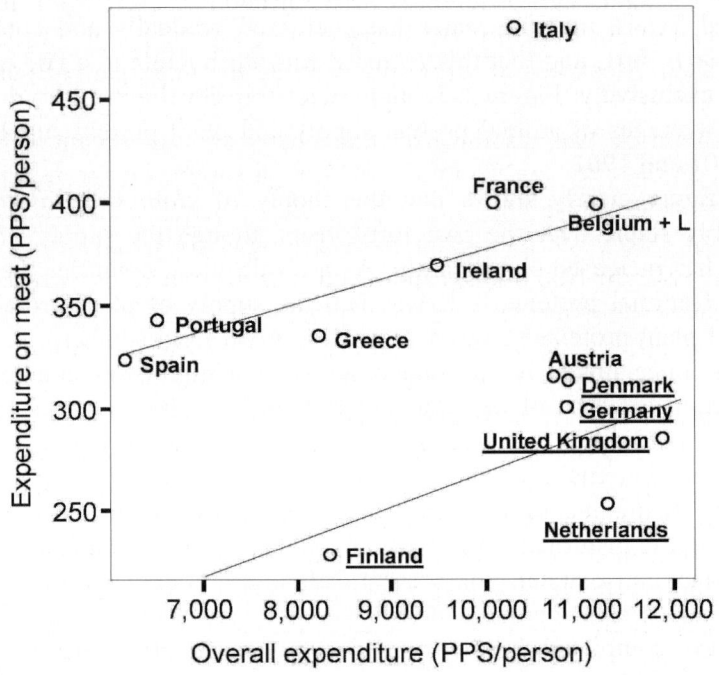

Figure 1-6. Expenditure on meat in 1999 as a function of annual household expenditure in Purchasing Power Standards (PPS) per person (Eurostat, 2005) among predominantly Protestant countries (names underlined; no data on Sweden) and non-Protestant countries.

Finally, these differences should not only be considered from the perspective of economy and culture, but also from a health perspective. However, this requires more insight into actual protein intake by specified cohorts. The Netherlands is one of the countries where elaborate data sets on actual consumption are available. This provides an opportunity to make a comparison between FAO data on supply and recent insights on a healthy diet. According to the most recent survey (Voedingscentrum, 1998), the average daily intake of protein in 1998 was 80 g per person, which is considerably lower than the supply of 106 g. The gap between the figures is primarily due to losses between national supply at retail level and actual consumption by the individual. The daily intake of 80 g is, on average, higher than RDI (recommended daily intake) values, which are specified in accordance with sex and age. Based on recent insights on a healthy diet, the over-consumption of protein is about 60%, which is not considered a health risk (Health Council of the Netherlands, 2001).

This brief protein consumption inventory of EU-15 provides some interesting insights into differences between national diets. As it turns out, there is a lot of diet diversity, but everywhere the same major sources can be distinguished. Meat, cereals and milk provide the main part of our dietary proteins. Interestingly, meat contributes only slightly more than cereals. As a general trend it was established that there are significant differences between, on the one hand, countries with high supplies of protein provided by cereals and vegetables, and, on the other hand, countries with high supplies of protein derived from milk. Portugal, Italy and Greece can be contrasted with the Netherlands and Sweden as the two poles of an axis, with intermediate positions for the ten remaining countries. A number of interrelated differences between these countries clearly demonstrated the impacts of path-dependent combinations of ecological, economic and cultural factors on current dietary protein supply.

Evidently, additional research is needed to specify more precisely the determinants of protein consumption. For example, the relationship between ecological, economic and cultural development often raises difficult questions on causes and effects. Nevertheless, the data depicted in Figures 1-4 to 1-6 may become of crucial importance for any attempt to design policy options that should contribute to more sustainable protein consumption. In considering options, path-dependent combinations of ecological, economic and cultural criteria will have to be taken into account. For example, in the Netherlands, which is displaying an extremely low level of plant protein combined with a quite high level of animal protein and a rather low level of household expenditure on food in general, the approach may have to be focused entirely differently from that in a Mediterranean country. It stands to reason that such will hold even stronger for continents other than Europe.

REFERENCES

Alexandratos, N. (1995), *World agriculture: Towards 2010 (An FAO study)*, J. Wiley & Sons, Chichester, UK.

Askegaard, S., and Madsen, T.K. (1998), "The local and the global: Exploring traits of homogeneity and heterogeneity in European food cultures", *International Business Review*, Vol. 7, pp. 549-568.

Baudet, H. (1986), *Een vertrouwde wereld: 100 jaar innovatie in Nederland,* Bert Bakker, Amsterdam, The Netherlands.

Bradford, E. (1999), *Animal agriculture and global food supply,* Available at <www.cast-science.org>, Council for Agricultural Science and Technology, Washington, DC.

Bréchon, P. (2004), "L'héritage chrétien de l'Europe occidentale: qu'en ont fait les nouvelles générations? (Christian heritage of western Europe: How have the new generations handled it?)", *Social Compass*, Vol. 51, pp. 203-219.

Brown, L.R. (1996), *Tough choices: Facing the challenge of food scarcity*, W.W. Norton & Co., New York.

Brundtland, G.H. (1987), *Our common future*, Oxford University Press, Oxford, UK, World Commission on Environment and Development.

CIA (2004), "The world fact book 2003", Available at <www.cia.gov/cia/publications/factbook/index.nl>, CIA, Washington, DC.

Cohen, M.D., March, J.G., and Olsen, J.P. (1972), "A garbage can model of organizational choice", *Administrative Science Quarterly*, Vol. 17, pp. 1-25.

Delgado, C., Rosegrant, M., Steinfeld, H., Ehui, S., and Courbois, C. (1999), *Livestock to 2020: The next food revolution*, Available at <www.ifpri.cgiar.org/index1.asp>, IFPRI /FAO/ILRI, Washington, DC.

Dosi, G. (1988), "Sources, procedures, and microeconomic effects of innovation", *Journal of Economic Literature*, Vol. 26, pp. 1120-1171.

European Communities (2001), *Consumers in Europe; facts and figures, data 1996-2000*, Office for Official Publications of the European Communities, Luxembourg.

Eurostat (2005), *Household budget surveys 1994 and 1999*, Statistical Office of the European Communities, last accessed 8 April 2005, Luxembourg.

Evans, L.T. (1998), *Feeding the ten billion: Plants and population growth*, Cambridge University Press, Cambridge.

FAO (2004), "Food balance sheets", (Last updated: 30 June 2003), Available at <apps.fao.org /page/collections?subset=nutrition>.

Fischler, C. (1999), "The 'McDonaldization' of culture", in Flandrin, J.L., Montanari, M., and Sonnenfeld, A. (Eds.), *Food: A culinary history from antiquity to the present*, Columbia University Press, New York, pp. 530-547.

Gilland, B. (2002), "World population and food supply: Can food production keep pace with population growth in the next half-century?", *Food Policy*, Vol. 27, pp. 47-63.

Grigg, D. (1995a), "The nutritional transition in Western Europe", *Journal of Historical Geography*, Vol. 21, pp. 247-261.

Grigg, D. (1995b), "The pattern of world protein consumption", *Geoforum*, Vol. 26 No. 1, pp. 1-17.

Health Council of the Netherlands (2001), *Dietary reference intakes: Energy, proteins, fats and digestible carbohydrates*, Available at <www.gr.nl> (in Dutch, with English summary), Health Council of the Netherlands, The Hague, Report 2001/19.

Hoffmann, R.C. (2001), "Frontier foods for Late Medieval consumers: Culture, economy, ecology", *Environment and History*, Vol. 7, pp. 131-167.

Ilmonen, K. (1991), "Change and stability in Finnish eating habits", in Fürst, E.L., Prättälä, R., Ekström, M., Holm, L., and Kjaernes, U. (Eds.), *Palatable worlds: Sociocultural food studies*, Solum Forlag, Oslo, Norway, pp. 169-184.

IMF (2004), "Per Capita Gross Domestic Product, Current Prices (U.S. Dollars per person)", (Last updated: April 2002), Available at <www.imf.org/external/pubs/ft/weo/2002/02/data /index.htm>.

Inglehart, R., and Baker, W.E. (2000), "Modernization, cultural change, and the persistence of traditional values", *American Sociological Review*, Vol. 65, pp. 19-51.

Keyzer, M.A., Merbis, M.D., Pavel, I.F.P.W., and Van Wesenbeeck, C.F.A. (2005), "Diet shift towards meat and the effects on cereal use: Can we feed the animals in 2030?", *Ecological Economics, (in press)*.

Kingdon, J.W. (1984), *Agendas, alternatives, and public policies,* Harper Collins, New York.

Kloppenburg, J., Lezberg, S., De Master, K., Stevenson, G.W., and Hendrickson, J. (2000), "Tasting food, tasting sustainability: Defining the attributes of an alternative food system with competent, ordinary people", *Human Organization*, Vol. 59 No. 2, pp. 177-186.

Langhelle, O. (2000), "Sustainable development and social justice: expanding the Rawlsian framework of global justice", *Environmental Values*, Vol. 9, pp. 295-323.

Lindblom, C.E. (1990), *Inquiry and change. The troubled attempt to understand and shape society*, Yale University Press, New Haven.

Millstone, E., and Lang, T. (2003), *The atlas of food: Who eats what, where and why*, Earthscan Publications Ltd, London.

Montanari, M. (1994), *The culture of food*, Blackwell, Oxford.

Nill, J. (2002), *Wann benötigt Umwelt(innovations)politik politische Zeitfenster? Zur Fruchtbarkeit und Anwendbarkeit von Kingdons "policy window"-Konzept*, Institute for Ecological Economy Research, Discussion Paper No. 54/02, Berlin.

Pezzey, J. (1992), *Sustainable development concepts: An economic analysis*, The World Bank, World Bank Environment Paper, number 2, Washington D.C.

Pimentel, D., and Pimentel, M. (2003), "Sustainability of meat-based and plant-based diets and the environment", *American Journal of Clinical Nutrition*, Vol. 78, pp. 660S-663S.

Seidl, A. (2000), "Economic issues and the diet and the distribution of environmental impact", *Ecological Economics*, Vol. 34, pp. 5-8.

Smil, V. (2000), *Feeding the world: A challenge for the twenty-first century*, MIT Press, Cambridge (Mass.).

Smil, V. (2002), *The Earth's biosphere: Evolution, dynamics, and change*, MIT Press, Cambridge (Mass.).

Tansey, G., and Worsley, T. (1995), *The food system: A guide*, Earthscan, London.

Teuteberg, H.J., and Flandrin, J.L. (1999), "The transformation of the European diet", in Flandrin, J.L., Montanari, M., and Sonnenfeld, A. (Eds.), *Food: A culinary history from antiquity to the present*, Columbia University Press, New York, pp. 442-456.

Tilman, D., Cassman, K.G., Matson, P.A., Naylor, R., and Polasky, S. (2002), "Agricultural sustainability and intensive production practices", *Nature*, Vol. 418, pp. 671-677.

Van Stuijvenberg, J.H. (1969), *Margarine: An economic, social and scientific history, 1869-1969 (Honderd jaar margarine 1869-1969)*, Liverpool University Press (Martinus Nijhoff), Liverpool (The Hague).

Van Wijk, J.J., and Rood, G.A. (2002), *Besluitvormingsmodellen in het transitieproces, toegepast op vegetarisch voedsel*, RIVM Report 550000004/2002, Bilthoven.

Vellinga, P., and Herb, N. (1999), *Industrial Transformation Science Plan*, IHDP, Bonn, Germany.

Voedingscentrum (1998), *Results of the Dutch food consumption survey (Zo eet Nederland 1998. Resultaten van de Voedselconsumptiepeiling 1997-1998, in Dutch)*, Voedingscentrum, The Hague.

Weaver, P., Jansen, L., Van Grootveld, G., Van Spiegel, E., and Vergragt, P. (2000), *Sustainable Technology Development*, Greenleaf Publishing Ltd, Sheffield (UK).

White, T. (2000), "Diet and the distribution of environmental impact", *Ecological Economics*, Vol. 34, pp. 145-153.

CHAPTER 2

ENVIRONMENTAL SUSTAINABILITY

Harry Aiking, Joop de Boer, Martine Helms,
David Niemeijer, Xueqin Zhu,
Ekko C. van Ierland & Rudolf S. de Groot

2.1 INTRODUCTION[1]

The concept of sustainability in general and food sustainability, in particular, entails many aspects and many interpretations (Aiking and De Boer, 2004). Extreme forms of over-consumption and wasteful management of natural resources are clearly unwanted, but there are strongly competing ideas about allegedly ideal sustainable production systems, such as organic agriculture and integrated crop management. In order to be able to claim that a certain food supply system is more sustainable than another, therefore, it is important to specify the sustainability aspects addressed and to develop sound methods for their assessment. Given the complexities in the case of proteins, it is necessary to first develop a number of theoretically consistent perspectives on the ecological basis of future protein supply. Subsequently, new tools are to be described concerning comparative evaluation of protein supply systems, which have been adapted to the needs of PROFETAS. Therefore, this chapter will focus on "environmental sustainability", but first place it in a broader context.

The sustainability of food supply systems can only be assessed comprehensively after specifying potentially relevant issues at a lower level of abstraction than the concept of sustainability itself. As a first approximation, Table 2-1 presents seven categories of potentially relevant aspects, namely global and regional environmental change, human health, human rights, labour conditions, animal welfare, and fair trade relations. It should be added here that animal welfare is a contested topic as an element of sustainable development. People in Northern European countries tend to see this as a matter of justice (because animals have the capacity to suffer), whereas people in Southern European countries hold a different opinion. At any rate, these categories encompass many of the topics that are incorporated

---

[1] Harry Aiking & Joop de Boer

in research on sustainable development (Bossel, 1996; Brundtland, 1987). For each category of sustainability aspects, Table 2-1 also shows policy issues at a lower level of abstraction. They represent "ills" (e.g. waste) or "ideals" (e.g. high biodiversity), which are considered relevant by stakeholders in society. The issues are based on distinctions made in relevant international documents, such as the 6th Environment Action Programme of the European Commission, the recommendations of the International Labour Organisation and the UN Declaration for Human Rights.

Table 2-1. Sustainability issues (*see the text) that are potentially relevant in the context of food (adapted from: De Boer et al. (2004)).

Aspects	Categories	Issues
Ecological aspects	Global change	Climate change
		Biodiversity
		Natural resources (land, freshwater)
	Regional change	Nutrient cycling
		Specific ecosystems
		Specific species
		Waste
Social aspects	Human health	Food security
		Food safety
	Human rights	Forced labour
		Child labour
		Non-discrimination
		Rights of indigenous people
		Freedom of association/bargaining
	Labour conditions	Wage levels
		Occupational health and safety standards
		Working hour
	Animal welfare*	Free-range
Economic aspects	Fair trade relations	Guaranteed price
		Long-term contracts
		Advanced payments and credit facilities
		Technical assistance
		Community support

Which categories of Table 2-1 should be specified and detailed depends on the production processes or products in question. What is at stake in a particular sector, such as cattle breeding or fishery, can in principle be specified in terms of "ills" or "ideals," although it will not be possible to express all issues in quantified indicators (e.g. respect for the rights of indigenous people). The issues relevant to sustainability may refer to each of the activities that are links in the supply of a specific food (e.g. beef or fish). More specifically, this amounts to the chain of activities, originating in the food's primary production, which are structured in a way that is distinct from that of other foods, even if they have certain components in common, such as transport or shops (e.g. Fine *et al.*, 1996).

As Table 2-1 demonstrates, the claim that a certain food supply system is more sustainable than another can only be substantiated if it satisfies a whole set of constraints. However, when a new system has to be compared with an old system, not all the issues may be applicable yet. This is a particularly relevant point, because the meat chain has been optimised for hundreds of years, but the plant-based NPF chain is still in its infancy and pea-derived NPFs are virtual products still. In fact, we have just begun to survey useful application of the non-protein fraction of potential crops (Section 3.7). On a geographic level, any impacts of the production and consumption of pea-derived NPFs on social aspects of sustainability do not only depend on the type of crops, but also on the location where they are grown.

An additional complication is that the issues mentioned in Table 2-1 are not completely independent of each other. First, there are dependencies through biophysical processes, for example, where climate change may have an impact on biodiversity. Second, there are dependencies through environmental impacts on social and economic processes, for instance, where climate change influences agriculture-related issues. Third, there are dependencies through the impacts of social and economic processes on the environment, such as the ecological footprints left by international trade, which may have impacts on both ecological aspects and on fair trade relations. These dependencies are not covered by conventional tools to compare the environmental performance of product chains, such as LCA (Life Cycle Analysis). Therefore, it is important to strive for improved quantification of performance.

Because we cannot address all these problems at once, this chapter focuses on the ecological aspects of Table 2-1, combining several ways to quantify environmental performance. This multi-method strategy, often called methodological triangulation, applies different methods to one research object. The idea underlying triangulation is that the relationships of the findings to one another may shed light on different aspects of a phenomenon (Erzberger and Prein, 1997). In other words, the related findings may converge and validate each other, they may stimulate new understanding by specifying complementary aspects, or they may show interesting divergences, which are difficult to explain. We will come back to these possibilities in the final section of this chapter.

The comparative analysis started off by specifying two model protein chains to develop a better understanding of the consecutive stages associated with a certain amount of human protein consumption (from primary production via processing and consumption to waste). As mentioned in

Chapter 1, the pork chain and the pea-based NPF chain were chosen as common points of reference for the calculations. Based on those chains, it is possible to calculate the conversion factors of proteins between the various stages and to make several comparative analyses of a (virtual) pea-based NPF chain and the (actual) pork chain in the Netherlands and the EU (Section 2.2).

Quantification of environmental performance was carried out in three PROFETAS projects, which took their starting points from economy, environment and ecology, respectively (Sections 2.3-2.5). They link economic activities to ecological effects, mediated by environmental emissions and land use allocation (Figure 2-1). The economic project began with a conventional LCA approach for gaseous emissions, complemented with a global economic model focusing on resource stocks. In the environmental project, the analysis started off by reviewing the ecosystem services that are the most relevant for comparing pork and pea-derived NPFs, indicating several options for aggregation of the results. The project on ecology set out to describe causal networks that highlight the interconnection of processes, specifying how indicator sets can provide a more complete picture of ecological effects.

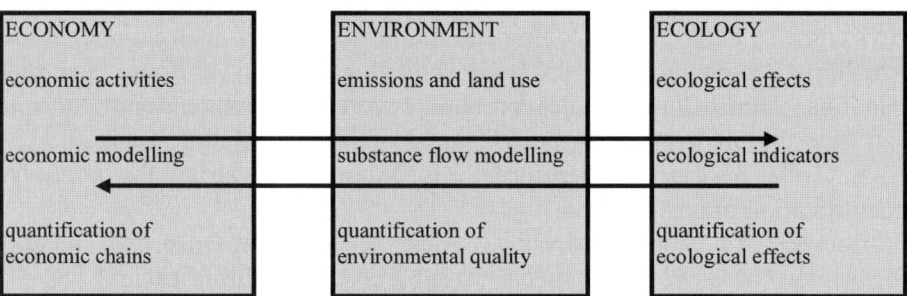

Figure 2-1. Framework of economic, environmental and ecological approaches to compare model protein chains.

2.2 THE PROTEIN CHAINS: PORK VS. PEA-BASED NPFs[2]

Two model protein chains were specified to compare pork products and pea-derived NPFs in terms of their environmental performance. The emphasis on model chains is in agreement with the PROFETAS philosophy that predicting actual products 20 years in advance is not feasible (and, in the case of the pork industry, carries the risk of getting lost in product details). Instead, it is better to develop tools to facilitate problem solving in the future. Moreover, because trying to mimic meat chops with plant proteins is not feasible, NPFs will have to serve as protein-containing diet ingredients, which can compete with meat-based ingredients in products such as composite meals, pizzas and snacks. For the sake of comparison, therefore, it is primarily necessary to examine how a given amount of protein for human consumption can be produced.

Both the pork and the NPF chain are derived in such a way as to deliver 1,000 kg protein for actual consumption (see Figure 2-2). The chains are basically composed of two important parts: agriculture and industry (consumption is not regarded as a separate entity here). The bottom part of the chains is mainly agricultural production (e.g. crop farming or pig farming) and the top part is mainly concerned with industrial production (meat processing or NPFs manufacturing) and consumption. Because NPFs replace meat ingredients, the NPFs are likely to show comparable characteristics with regard to cooking, storage etc., especially since the meat/NPF ingredient is just one of a variety of ingredients that constitute the final product at the retail and consumer stages. For this reason the production chains are assumed to be negligibly different above the stage of product processing (represented by the boxes named NPF and meat in Figure 2-2), and the parts played by retail and the consumer are therefore relatively unimportant. The pork production system represents an elementary supply chain mostly based on general Dutch production figures. The feed ingredients are simplified and chosen to represent primary agricultural products (tapioca) and processed products of other food industries (soy cakes). The NPF production system supplies a virtual product modelled after Quorn^{TM} minced and chopped meat. Proteins are isolated from peas by air classification (see Section 3.1) and the protein fraction consequently also contains a certain amount of carbohydrates and other pea components. The emphasis on the production chains presented in Figure 2-2 is on the inputs

[2] Harry Aiking, Martine Helms, David Niemeijer & Xueqin Zhu

and outputs of land, water, nitrogen and phosphorus. Additional attention is paid to the production of by-products and wastes, and to spilling of food that, although occurring in every stage of the chains, is accounted for as harvest losses and combined losses from retail to plate.

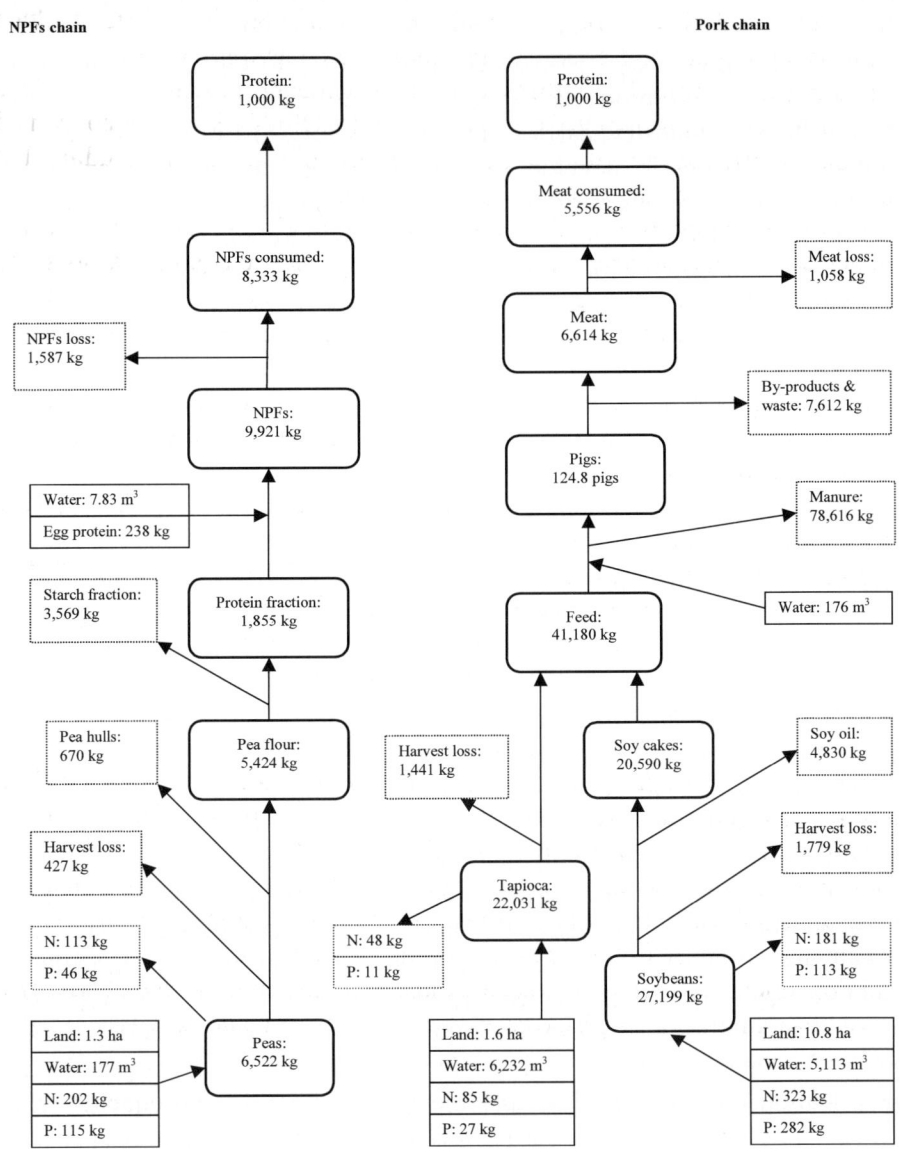

Figure 2-2. Simplified Dutch pork and pea-NPF production systems, resulting in comparable protein consumption.

The impacts of the protein food systems provided in Figure 2-2 have to be organised in a way that enables comparison. One way to do this is by means of indicators. Environmental and ecological indicators (NRC, 2000: 19) are increasingly used at all levels of society (Jackson *et al.*, 2000) to provide information on the pressures on the environment, the evolving state of the environment, and the existence of appropriate policy responses (Niemeijer and De Groot, 2005a). Environmental indicators include physical, biological and chemical indicators (Smeets and Weterings, 1999: 6) and generally comprise indicators of environmental pressures, conditions and (societal) responses (OECD, 1993: 6). In PROFETAS we are concerned with environmental indicators in the widest possible sense, so including, but not limited to, indicators of ecological processes.

From the perspective of environmental sustainability, the most relevant impacts of the agricultural bottom part of the chain include habitat loss and degradation through the use of land, soil nutrients and water and the emissions of nutrients, herbicides, pesticides, greenhouse gases (e.g. CO_2, CH_4 and N_2O) and other pollutants (e.g. NH_3). The top part of the chain, which concerns product processing, retail and consumption, brings about energy use, waste and emissions of air pollutants (such as SO_2 and NO_x) and greenhouse gases (such as CO_2, CH_4 and N_2O).

Comparing the inputs and outputs in Figure 2-2, the pea-derived NPF chain seems to score considerably better on environmental sustainability than the pork chain. This is mainly due to the inefficient conversion of plant protein to meat protein requiring large inputs in the agricultural phase to produce feed. However, both chains also produce non-protein output. For example, the by-products of the pig chain, primarily generated during slaughter (blood, hair and hooves, intestines, bones, skins), are almost completely processed further to valuable material, such as pharmaceuticals, cosmetics, fat, sausage casings, felting, leather, gelatine, phosphate fertiliser and glues. In contrast, useful application of the non-protein fraction (to be released during processing of potential crops for NPFs) is still under consideration (see Section 3.7). Both the economic feasibility and the environmental impact of NPFs production depend to a large extent on (the applications that may be found for) these non-protein by-products. In addition, energy input is not presented in Figure 2-2, but the use of energy in the NPF-manufacturing phase is expected to be considerable (see Section 3.6).

2.3 ECONOMIC APPROACH TO ENVIRONMENTAL SUSTAINABILITY OF PROTEIN FOODS[3]

2.3.1 Introduction

Comparing the pork and the NPF chain from the perspective of environmental economics aims to provide more insight into the co-evolution of economic and environmental processes. This refers to (a) the impacts of social and economic processes on the environment, (b) environmental impacts on social and economic processes, and (c) societal responses to these impacts. For example, a shift from meat to plant-based protein foods can be considered a societal response to the prospect of environmental degradation.

In order to identify solutions to environmental problems associated with protein foods, the above interactions can be studied by a combination of environmental assessment and economic modelling. Systems analysis of protein chains by means of Life Cycle Analysis (LCA) can be used for a preliminary assessment of their environmental impacts. Economic modelling of the environmental problems related to protein production and consumption can then be used to assess both the environmental and economic impacts of NPFs, which may identify options leading to more sustainable food consumption.

2.3.2 From LCA to economic modelling

LCA is an established tool to compare the environmental performance of production chains. If the methodological limitations of LCA are taken into account, it can be used as a first approximation of the differences (also see the following sections). An important limitation is that LCA is based on emissions, rather than on environmental effects, and that contextual factors, such as the climate and soil properties of a geographic location, are not addressed. Furthermore, its incorporation of sustainability aspects is rather arbitrary and it is unable to handle anything but existing products. In fact, LCA is intended for comparative use, i.e. the results of LCA studies have a comparative significance rather than providing absolute values on the environmental impact related to the product.

---

[3] Xueqin Zhu & Ekko C. van Ierland

At any rate, environmental pressure derives from either pollution or resource use. In LCA this is expressed by means of emission indicators and resource use indicators, respectively. Many environmental pollution problems caused by the pork chain are the result of an intensive production system and its manure surplus, as well as a poor conversion ratio from plant proteins to meat proteins. The large amount of minerals in manure affects the quality of soil, water and air. Considering the diversity of the emissions and their environmental impacts, emission indicators were based on conventional "environmental issues", because many emissions have comparable effects on the environment. The following five emission indicators were selected: CO_2 equivalents for global warming, NH_3 equivalents for acidification, N equivalents for eutrophication, pesticide use (kg active ingredient) and fertiliser use (kg). Since agriculture requires land and water as inputs, land use and water use indicators were selected to reflect resource use.

Under these assumptions, a preliminary LCA (Table 2-2) shows that under conditions prevailing in the Netherlands, the pork chain contributes 61 times more to acidification than the NPFs chain, 6.4 times more to global warming, and 6.0 times more to eutrophication. The Dutch pork chain also needs 3.3 times more fertiliser, 1.6 times more pesticides, 3.3 times more water and 2.8 times more land than the NPFs chain (Zhu and Van Ierland, 2003b). These rather conservative estimates (see Section 2.6) provided some of the building blocks to the more detailed analysis depicted in Figure 2-2.

Table 2-2. Preliminary emission and resource use indicators per functional unit (Source: Zhu and Van Ierland (2003b)).

	Pork	**NPFs**	**Pork/NPFs**
Acidification (NH_3 equivalent, kg)	675	11	61.0
Global warming (CO_2 equivalent, kg)	77,883	12,236	6.4
Eutrophication (N equivalent, kg)	2,491	417	6.0
Pesticide use (active ingredient, kg)	18	11	1.6
Fertiliser use (N + P_2O_5, kg)	485	144	3.4
Water use (m^3)	36,152	10,912	3.3
Land use (ha)	5.5	1.95	2.8

In order to add a number of relevant contextual factors to the analysis, it is necessary to consider the problems related to pork production in the Netherlands and the European Union. It is well known that the problems related to the Dutch pig production system are not only local but also global as a result of large-scale imports of feed. Feed production is quite land-intensive, which imposes a huge pressure on land in the developing world. For example, the increased production of raw materials for animal feed has resulted in large-scale deforestation in Thailand, Brazil and Argentina.

One of the options economic actors may consider is to locate more pig production in areas with arable products where transport costs of feed would be relatively low, and few problems would arise in terms of air, water and soil pollution. Agriculture is, however, often the economic locomotive of a region and an important source of direct and indirect employment. If livestock were simply reduced by five million pigs in the Netherlands, for example, it would mean a loss of 28,000 jobs (Bolsius and Frouws, 1996). Clearly, closing down pork production incurs economic costs. Thus, society will make trade-offs between the welfare effects of environmental degradation and economic welfare aiming to improve income.

An additional question is how the rest of world will react if EU consumers partly replace pork by NPFs. As far as pork is concerned, the problems are related to (a) the increased demand for meat in rapidly industrialising countries and (b) feed trade. As long as pork is highly demanded in the whole world and feed is imported from the rest of world, the pork issue is an international issue. For example, if the rest of the world, mainly some Eastern Asian countries, are to display an increasing demand for meat (Keyzer *et al.*, 2003), will the meat producers in the EU neglect it and only consider the EU market? To answer this question, we need a more dynamic analysis of the ways in which agricultural producers and traders may allocate their resources.

2.3.3   Environmental economic modelling

Economic modelling aims to provide a picture of how the whole economy – including the rest of the world and possibly its environmental state – will change as a result of the introduction of NPFs into the EU. For this study Applied General Equilibrium (AGE) models were chosen because they are suitable for studying world-wide issues (Shoven and Whalley, 1992; Ginsburgh and Keyzer, 1997). An AGE model with an environmental dimension was constructed for this study to analyse the implications of NPFs introduction for the economy.

The model that considers interactions between the economic system and the environmental system is a so-called ecological economics model (Costanza *et al.*, 2000). To include environmental aspects in an economic model several specifications are required. For an economic system consisting of production and consumption, environmental resources function as input, whereas substances emitted to the environment are the output. In the environmental system, resource stocks and emission inflows from economic activities change the quality of the environment following the laws of biophysical processes. The environmental quality supplies feedback to the economic system by influencing the amenity of the environment and economic productivity.

According to Fonk and Hamstra (1995), a substantial shift from pork to NPFs by consumers would imply that by 2035 the share of NPFs might have increased to 40% of meat expenditure. The model simulated the impacts of such a shift. Unsurprisingly, the results show that pork production and consumption in the EU will decrease with the increase of NPFs but the reduction in pork production is smaller than the reduction in pork consumption as long as pork is still demanded in the rest of the world (Zhu et al., 2004).

Pork production is partly dependent on consumers' concerns for environment, health and animal welfare, as well as price and income concerns. These issues were translated into the economic model by a parameter for the utility attributed to the environment and the substitution between pork and NPFs in budget choices. In this case, the results of the model show that a higher concern for environmental quality increases production and consumption of NPFs and decreases pork consumption in the EU. Interestingly, the results were more sensitive to the parameter for environmental concern than to the parameter for substitutions between pork and NPFs. In other words, the political role of consumers (as citizens) who support more stringent environmental standards might be greater than their economic role when they make trade-offs between the prices of pork and NPFs. It should be noted that this is a remarkable, but true, result and not an artefact that can be attributed to the assumptions of the model.

A final item in the present analysis was the heterogeneity of environmental concern in the world. Higher environmental concerns in the EU compared to the rest of the world results in more emissions in the region with lower values of environmental concerns. Unsurprisingly, it can be said that NPFs are meaningful for *global* environmental improvement only if both regions increase their preferences for environmental quality (Zhu and Van Ierland, 2003a).

2.3.4 Conclusions

In summary, it can be concluded that NPFs are clearly environmentally more friendly than pork, but also that the real impact depends on consumer acceptance of NPFs. At any rate, it seems likely that the European production of pork will continue for economic reasons. Even if many European consumers switch to consume more NPFs to replace a substantial part of pork, pork production in the EU might not be reduced because of higher demand of meat in rapidly industrialising countries (see Section 5.3).

Whether it is profitable to continue pork production for economic reasons is partly dependent on the environmental standards set by governmental policymakers. It is particularly because of this linkage that the political role of consumers who support more stringent environmental standards might be greater than their economic role when they make trade-offs between the prices of pork and NPFs (the citizen-consumer issue).

Differences between the level of environmental concern in the world's major regions may result in more pollution in the regions with less environmental concern. This means that local transition from meat protein to plant-based protein foods might not achieve more environmental sustainability on a global level, because more meat might be consumed elsewhere, in particular, in Asia and in South America. Accordingly, North-South issues are emerging, i.e. the large differences between developing and developed countries are more important to the overall PROFETAS goals than originally realised (see Chapter 6).

2.4 MEASURING ENVIRONMENTAL SUSTAINABILITY OF PROTEIN FOODS[4]

2.4.1 Introduction

One of the main questions from the perspective of environmental sustainability assessment is how to find a systematic way to select the elements and processes that are relevant for a comparison of pork and pea-derived NPFs. The option chosen in this section is to start off from the reliance of human society on the benefits that the environment provides and to focus on ecosystem services. These benefits are embodied by goods such as oxygen, oil, fish or timber and services such as climate regulation, soil formation and enjoyment of landscape scenery, e.g. see De Groot *et al.* (2002). Taken together, these benefits are called the ecosystem services and they are generated by physical ecosystem properties and processes that are often referred to as ecosystem functions (Costanza *et al.*, 1997; De Groot *et al.*, 2002). The nature and extent of present human activities disrupt the ecosystem processes, reducing the capacity of the environment to consistently yield ecosystem services and consequently threaten long-term human welfare.

[4]Martine Helms

2.4.2 Approach

For the purpose of this section it will be assumed that environmental sustainability refers to safeguarding future flows of ecosystem services and therefore to maintaining the functions that yield these services. The reliance of protein food production on ecosystem services, the damage that protein food supply does to the ecosystem functions that support the services, and the feedback to the future service generating capacity are at the heart of the present methodology developed.

The main issues in this assessment are loss of ecosystem functions and service generating capacity. Qualitative relationships between food production and required services, on the one hand, and food production and function damage, on the other hand, are or can be established.

In general, food production requires land, water, nutrients, energy, stable climatic conditions, genetic exchange and diversity to provide some resilience against pests and adverse conditions. Therefore, it requires services such as climate regulation, soil formation and pollination. At the same time, food production disturbs services such as:

- refugium and genetic resources, by habitat and biodiversity loss, due to the cultivation of natural areas and eutrophication,
- climate regulation, due to changes in land use and the release of greenhouse gases by fertilisation, livestock keeping and the use of fossil fuels, and
- water supply, due to large-scale irrigation.

Relationships between damage, feedback mechanism and future service generation are not easily established. In addition, quantification of the impact of food production on ecosystem functioning and the provision of nature's services provide some difficulties. It is possible to determine the impact of food production on the processes that lead to ecosystem changes, which are the "conventional" environmental issues such as greenhouse effect, acid rain and eutrophication. Although these issues do not inform us on the extent to which food production influences environmental sustainability, establishing potential links between ecosystem functions and conventional environmental issues may provide a promising approach to estimate function loss (Figure 2-3).

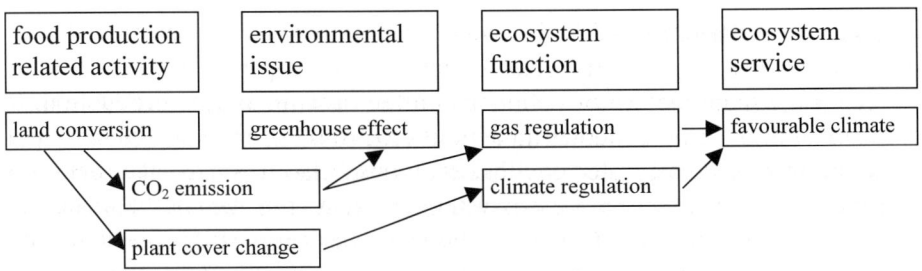

Figure 2-3. Example of links between food producing activities, conventional environmental issues and ecosystem function losses.

The conventional environmental issues found to be important in protein production (Helms and Aiking, 2003; Helms, 2004) include (a) biodiversity loss (mainly due to land conversion), in addition to impacts on (b) the water cycle, (c) the carbon cycle (climate change), (d) the nitrogen cycle, (e) the phosphate cycle. These issues were thought relevant for comparing different production systems and thus served as a basis for function selection. Most ecosystem functions represent processes that are internal to the global ecosystem and essential to its overall functioning. It is assumed that the working capacity of the ecosystem processes that is diverted by human activities is no longer of benefit to the "natural" part of the ecosystem. The functioning of the system is then disrupted in a measure equal to the amount of working capacity that is diverted. For that reason the appropriation of functions for human activities (here, protein production) is held as a measure of the impact on environmental sustainability.

In a first application of the method six functions are chosen. These include three enabling functions facilitating the protein production system: (1) biomass production, (2) energy production, (3) water supply and three remedial functions that prevent impacts from occurring or that restore the damage after impacts have occurred, (4) biodiversity conservation, (5) nitrogen retention, and (6) phosphate retention.

Once function appropriation has been estimated, it will have to be valued to enable comparison between the effect of different production systems. Presently proposed units of aggregate expression are temporal (time required for a natural recovery) or spatial (area required). The area use assessment focuses on the amount of productive area required for the functions to take place during a length of time relevant to the production process (set at the production of annual crops, or about 1 year). The time frame assessment focuses on the time that is required for the functions to take place in a limited area.

The appropriated functions are described as follows:

Biomass production – The impact of biomass production is calculated as the area or time required to yield the amount of crop product that is used either for the manufacture of NPFs or as a feedstuff in animal production. It is assumed that a crop occupies an area of cropland for one year, so that land does not need to be allocated to different crops. Also it is assumed that crop and bio-energy production are continuous so that they can be harvested at any time during the year. Generally, however, a crop occupies a field only part of the year and can only be harvested at one point in time (batch culture).

Energy production – The time or area required to produce a substitute biofuel is used as a measure of the environmental impact of the energy used in the protein production chains (for synthetic nitrogen fertilisers as well as for processing). Bio-energy is a renewable energy source with the additional advantage that its use avoids further CO_2 accumulation in the atmosphere. It is estimated that annually about 300 GJ per hectare can be produced in the form of biomass in the Netherlands (Van Doorn, 1993). The most efficient N-synthesis plants use 27 GJ for the production of 1 ton NH_3 or 32.8 MJ for 1 kg nitrogen (Smil, 2001).

Water supply – The impact of water use is expressed as the area or time required to receive an amount of precipitation (rain) equal to what is used in the irrigation of crops and production of pork. It is implicitly assumed that precipitation is equally spread over the year and over the country's area, and that all precipitation is retained as a (virtual) water resource of which none is lost due to evaporation or use and transpiration by plants. Data used are derived from the FAO (2004) for precipitation and from the WRI (2004) for total land area.

Biodiversity conservation– The reservation of natural area is used to mitigate the impact of land cultivation on biodiversity. It is assumed that 35% of species is required for general ecosystem functioning and that in order to save 35% of species, 35% of natural area must be conserved. Therefore, for every 0.65 ha that are required for biomass or energy production, 0.35 ha of natural area are set aside for conservation purposes. To preserve biodiversity well, these areas will have to remain largely undisturbed (natural) for prolonged periods of time. For the moment no satisfactory biodiversity reservation conversion ratio to time has been found. As a result, in the time frame method, biodiversity reservation is omitted.

Nitrogen and phosphate – The environmental impact of nitrogen and phosphorus emissions, for example due to the use of fertiliser in the crop production phase and the excretion of manure in the animal production phase, is assessed using the nutrient assimilation capacity of Dutch wetlands. Assuming that they consist of equal areas of reed, peat, open water and forest, the average N-retention rate is 119 kg/ha.yr and the average P-retention rate is 7 kg/ha.yr (Goosen *et al.*, 2004). Manure is considered a waste product that is disposed of by using it as fertiliser and all emission impacts are allocated to the animal production. Crop plants do not efficiently use the nutrients in the fertiliser that is applied. It is assumed that the crop uses 44% of the applied nitrogen and 56% is lost to the environment – 8% to soil, 31% to water and 17% to air. Of the applied phosphorus it is assumed that plants use 60%, the soil absorbs 28% and nearby water bodies receive 12% (Baan and Hopstaken, 1989; Carpenter *et al.*, 1998).

2.4.3 Practical application of the instrument to pork and to pea-NPFs

The result of a first application of the area assessment method to the pork and pea-NPF production system as described in Section 2.2 is shown in Table 2-3. The area required to support the production of 1,000 kg protein is 26.2 ha for pork and 4.6 ha for pea-NPF.

Table 2-3. The area (ha) required to fulfil the functions appropriated for protein production.

Function	Pork	NPF	Area type
Crop production	10.27	1.35	Production/crop
Energy production	0.04	0.02	Production/energy
Water supply	0.95	0.02	Natural/wetland
Biodiversity conservation	5.55	0.74	Natural/all
N-retention	2.63	0.53	Natural/wetland
P-retention	6.73	1.97	Natural/wetland

However, it is safe to assume that several functions are performed simultaneously and do not interfere with each other. Therefore, water supply, N- and P-retention are appointed the same wetland area – that is the largest area use of the three functions (in this case, that of P-retention). In this *prudent-function-combination* assessment the area appropriation of pork is 22.6 ha and that of NPF is 4.1 ha (Table 2-4). If the three natural area services are combined in a *maximum-function-combination* assessment, the appropriation of pork is 17 ha and the appropriation of NPF is 3.3 ha (Table 2-4).

Table 2-4. The effect of function combinations and area differentiation on area (ha).

Assumptions	Pork	NPF
No function combinations (NFC)	26.2	4.6
Prudent function combination (PFC)	22.6	4.1
Maximum function combination (MFC)	17	3.3

2.4.4 The (partial) replacement of pork with pea-NPFs

The hypothesis of the PROFETAS research programme that is at the basis of this study is that a partial replacement of animal proteins with plant proteins is environmentally more sustainable. In the preceding section the difference in impaired environmental sustainability was calculated for the production of 1000 kg protein. In this section the difference is described if part (0% – 20% – 50% – 80% – 100%) of the pork protein in Dutch and European diets is replaced with pea-NPFs in the year 2004, using the prudent-function-combination for the area assessment. For the Netherlands it is assumed that 16.3 million Dutch people will each consume 48.6 kg pork on average. This amounts to a total of 792 million kg pork, or 143 million kg pork proteins. For EU-15 it is assumed that the 380.4 million inhabitants consume 44.4 kg pork, on average. This amounts to a total of 16.9 billion kg pork or 3.04 billion kg pork protein to be partially or wholly replaced. These results are presented in Table 2-5.

Table 2-5. Protein-production area required when replacing pork with NPF.

Scenario	Pork (%)	NPF (%)	area NL (ha)	area EU-15 (ha)
1	100	0	3,230,000	68,685,000
2	80	20	2,701,000	57,426,000
3	50	50	1,907,000	40,537,000
4	20	80	1,112,000	23,649,000
5	0	100	583,000	12,390,000

2.4.5 Discussion and conclusions

In this section the environmental sustainability of protein production is measured in terms of ecosystem functions that are deployed to support protein production or remedy the ecological effects, expressed on an area basis. Depending on the combination of functions, the area used for the production of pork protein is 5.2-5.7 times larger than the area used for pea-NPF proteins. Based on this assessment it can be concluded that the production of NPFs is five and a half times more sustainable than the production of pork. Of the six ecosystem functions that are assessed, the functions responsible for growing crops, mitigation of P-outputs and moderating biodiversity loss account for most of the impact on environmental sustainability.

Up to now, the analysis mainly covers the agricultural and animal production phases. The impacts of the remaining parts of the production process, i.e. crop processing, transport, storage, retail, consumption, waste, are expected to be relatively small with respect to land use, water use, and P-emissions, but in comparison to the production phase the energy use in the processing part of the production chains may still be considerable (Carlsson-Kanyama, 1998). These phases cannot be ignored, therefore, and have to be added in the final version of the methodology. In addition, a way is sought to incorporate possible positive impacts of protein production such as the N-fixation potential of leguminous plants to soil fertility, and the destiny of the 70-80% non-protein fraction of peas should be taken into account. The latter, however, is the subject of a dedicated research project (see Sections 3.7 and 6.3).

2.5 ECOLOGICAL INDICATORS FOR SUSTAINABLE FOOD PRODUCTION[5]

2.5.1 Introduction

Indicators provide a viable means to assess the environmental performance of food production and consumption and to make comparisons between different chains. The existing frameworks of indicators may become more useful if they incorporate more ecological process knowledge. This means that an approach has to be developed for the coupling of indicators of resource use and emissions by food production with indicators for the effects or impacts on the ecosystems in which the production takes place. Such an approach allows us to make use of environmental process knowledge without requiring the detail and the large amounts of data necessary for an adequate numerical simulation or a full life cycle analysis (LCA). Process knowledge provides insight into cause-and-effect networks and can guide the selection of appropriate indicator sets.

2.5.2 Approach

Two approaches were developed: (1) a theoretical approach was developed in the form of a method to select an appropriate set of indicators that would reflect the causal relations between driving forces and pressures of food production, on the one hand, and environmental state and impacts, on the other hand; (2) an applied approach was developed that permitted a

---

[5] David Niemeijer & Rudolf S. de Groot

qualitative conversion between environmental pressures and affected ecosystem functions.

The theoretical approach. Environmental indicators are usually considered within the context of one of three popular frameworks: Pressure – State – Response (PSR), Driving force – State – Response (DSR) and Driving force – Pressure – State – Impact – Response (DPSIR) (Hammond *et al.*, 1995; OECD, 1999; Smeets and Weterings, 1999). *Driving forces* typically refer to social and economic developments that indirectly affect the environment. *Pressures* concern emissions, resource use and other processes that exert a pressure on the environment. *State* involves the changes in the biological, chemical or physical state of the environment. *Impact* refers to the effects such changes have on human health and ecosystems. *Responses*, finally, refer to the societal responses to all these changes in the form of management actions or policies.

Getting from pressure indicators to environmental consequences of those pressures is an important step. Traditionally, this is done by selecting an indicator for each combination of an environmental issue (e.g. eutrophication, ozone depletion, global warming) and a step in the causal chain (e.g. pressure, state, impact). This produces a matrix of indicators with environmental issues on one axis and the causal chain on the other axis. The selection process is mostly done on the basis of expert opinion and criteria applied to indicators individually. Indicator studies tend to be focused on finding the indicators for which the most data is available or that on theoretical grounds best reflect a particular pressure, state or response (Niemeijer, 2002).

During a literature review, no framework or approach turned out to be available for developing a *coherent set of indicators*, which (1) links pressures to environmental consequences and (2) focuses on how each of the indicators can contribute to the problem solving logic needed to answer questions of environmental sustainability. The existing PSR, DSR and DPSIR frameworks do emphasize the fact that processes and therefore indicators are causally related. However, they tend to simplify these interrelations to a series of individual causal chains of the type: halocarbon emissions (pressure) affect the ozone column (state) and CFC recovery programs (response) try to reduce the pressure. While the causality of such a chain is obvious, it is a very limited view of the real world in which, for example, a reduction in the ozone column in turn leads to changes in UV radiation, which in turn affects biological life forms and therefore species abundance, which is considered part of another causal chain focused on biodiversity (Niemeijer and De Groot, 2005b).

Because the existing frameworks take a rather one-dimensional perspective, the process of indicator selection tends to be reduced to finding the best driving force, pressure, state, impact and response indicators for each environmental issue of concern. It has been argued (Niemeijer and De Groot, 2005a; Niemeijer and De Groot, 2005b) that by working with *causal networks*, which highlight the multidimensional interconnections of processes, much more insight can be gained into the interrelation of indicators and therefore how individual indicators can contribute to providing a more complete picture of what is happening in the environment. This provides essentially an enhanced DPSIR (eDPSIR) framework. Constructing a causal network for a particular problem domain can help locate the key nodes for which we need indicators. The existing individual indicator selection criteria can then be applied to select the best indicators for these key nodes (Niemeijer and De Groot, 2005a). Such an approach makes the best use of the wealth of process knowledge that we already possess on the linkages between food production and the environment without having to develop a full-fledged numerical model with all its complications (Niemeijer, 2001).

The applied approach. The approach outlined above works well if sufficient data is available for the different DPSIR categories. However, data on pressure indicators for food production is generally much more readily available than on any of the other DPISR categories. This is even more an issue if we want to trace the global impact of pork and pea production in the Netherlands. For that reason a more applied approach was developed. This approach uses the concept of ecosystem functions (De Groot, 1992) to translate the pressure indicators to actual environmental impacts. Ecosystem functions may be defined as "the capacity of natural processes and components to provide goods and services that satisfy human needs, directly or indirectly" (De Groot, 1992: 7). In this approach the main sources of environmental pressures are identified (covering resource use and emissions). For each of these pressure sources, key indicators are identified that can be applied to a wide range of processes in the food production chain. Next, a spatial breakdown is established of the main steps in the production chains and the key environmental pressure indicators are quantified for each country. Finally, for each selected pressure indicator it can be determined which ecosystem functions are affected and whether these functions are lightly or strongly impacted. The more functions a pressure affects the stronger the total impact on ecosystems will be. This allows a qualitative assessment of the spatial distribution of environmental impacts of pork and pea production in the Netherlands (Niemeijer *et al.*, 2005).

2.5.3 Application to the pork and NPFs chains

An analysis of the pork and pea-NPFs chains is dealt with here according to the above approach (Niemeijer *et al.*, 2005). The focus of this analysis is on the production part of the chain and not on the consumption part, as the pressure on the environment of consumption and consumer waste generation is likely to be small compared to the pressure exerted during the various production steps. Moreover, differences between the two chains can be expected to be relatively small for the consumption component. Our analysis of production chains will be limited to primary production and exclude industrial processing of meat and peas to raw proteins. There are two reasons for this focus. First, the largest difference in environmental burden between the two chains can be expected in the primary production phase, in particular, as a consequence of resource depletion, since over one-third of all ice-free land area is used for food production, along with over three quarters of the available freshwater (Smil, 2002: 239). Second, as pea-NPFs are at present only hypothetical products, insufficient information is available on the processing procedures to allow for an adequate indicator-based comparison with meat-based products.

Key pressure indicators were selected for resource use and for emissions. For resource use these were: Area used (ha), Energy used (MJ), Soil nutrient extraction (N-equivalents in kg) and Volume of water used (m^3). For emissions these were: Eutrophication (N-equivalents in kg), Acidification (NH_3-equivalents in kg) and Global Warming (CO_2-equivalents in kg). Together, these encompass the most important sources of environmental pressure across a wide range of production processes and they are also indicative of the effect on key geochemical cycles. The analysis involved determining a spatial breakdown for these indicators for fertiliser production, plant production, transport and pig production. For the pork chain, production board data was used to estimate how much of each of the ingredients of pork feed is derived from each contributing country. Ingredients for pork feed were found to be derived from over 15 different countries spread over five continents. Pea production was assumed to take place entirely in the Netherlands. In summary, it was found that pea-NPFs production scored 4 to 200 times better than pork production (depending on the indicator). Table 2-6 provides an overview of the differences. It should be kept in mind, however, that the energy involved in processing has not yet been included and that the energy indicator may, therefore, yield an over-optimistic view.

Table 2-6. Differences in environmental pressures between the pork and pea chains for a functional unit of 1,000 kg protein (* soil acidification caused by fertiliser use not included).

	Pork	Pea-NPFs	Pork/NPFs
Resource use indicators			
Area (ha)	7.6	0.8	9.5
Energy (MJ)	196,000	1,600	123
Water use (m^3)	40,000	9,300	4.3
Nutrient depletion (kg N-equiv.)	4,400	440	10.0
Emission indicators			
Eutrophication (kg N-equiv.)	3,000	210	14.3
Acidification* (kg NH_3-equiv.)	375	2	192
Global warming (kg CO_2-equiv.)	30,000	270	111

Results such as those presented in Table 2-6 could also have been derived from an LCA. What a typical LCA would not reveal is where the environmental burden is placed. Table 2-7 shows for pork production where the resources are used and where the emissions occur. It is a compilation of the country level data used for the analysis. In the table it can be seen how pork production in the Netherlands mainly uses natural resources outside of the country, with a 30 to 40% share for developing countries. Even for the emissions, about half the environmental burden occurs outside the country (inside the Netherlands the emissions are mainly related to manure production by the pigs). Because of the low input production methods used in many of the developing countries, their share of the emissions burden is low compared to their resource use share. To some degree this can be considered positive, but it is also indicative of the considerable transport of nutrients from developing countries (where they are not sufficiently replaced with fertiliser) to the Netherlands where they create excessive environmental emissions through manure.

Table 2-7. The share (%) of the global environmental pressures of Dutch pork production.

	Resource use			Emissions		
	Area use	Water use	Nutrient depletion	Eutrophication	Acidification	Global warming
Developing countries	39	27	28	8	4	5
Industrialized countries	61	72	71	46	20	31
The Netherlands	0	1	1	46	62	47
Oceans and seas	0	0	0	0	15	17
Total	100	100	100	100	100	100

Going beyond environmental pressure towards the impact on ecosystem functions, our qualitative analysis revealed, as could be expected, that the effect of pork production was much stronger than that of pea production. The most notable difference involved the ecosystem production function. Looking again at the spatial dimension, the regulation, habitat and carrier functions were most affected in the industrialized countries, with the developing countries and the Netherlands more or less in a shared second place. The production functions were most strongly affected in the Netherlands, followed by the other industrialized countries. The information function was seriously affected in the industrialized and developing countries and hardly at all in the Netherlands where pork production occupies relatively little space. Also the ecosystem functions of the oceans and seas where a lot of the transport takes place were negatively affected in all cases except the information functions.

2.5.4 Conclusions

Indicators provide a viable means to assess the environmental impacts of food production and consumption and provide one way to make comparisons between different food production and consumption chains. A theoretical approach was developed in the form of a method to select an appropriate set of indicators that would reflect the causal relations between driving forces and pressures of food production, on the one hand, and environmental state and impacts, on the other hand. An applied approach was developed that permitted a qualitative conversion between environmental pressures and affected ecosystem functions. Working with indicators allows us to make use of environmental process knowledge without the detail and the huge amounts of data requirements for good numerical simulation or a full life cycle analysis (LCA). Most importantly, however, process knowledge provides insight into cause-and-effect networks and guided the selection of appropriate sets of indicators.

The applied approach allowed us to get insight into the spatial dimension of the Dutch pork and pea-NPFs production chains without the massive amounts of data required for a more traditional approach, which are also difficult to obtain. It was found that to produce the same amount of raw protein, pork production exerts – depending on the indicator used – 4 to 200 times higher pressures on the environment than pea production. However, it should be emphasised that energy-related indicators are over-estimated, because only the primary production of peas have been taken into account. To a very large extent these higher pressures for the pig chain have their impacts outside the Netherlands. By linking pressure indicators to ecosystem functions it was possible to make a first approximation of the impact of the pressures exerted by the pork production chains on ecosystem functions in different parts of the world. The ecosystem production function is affected the most in the Netherlands, whereas the regulation, habitat, and carrier functions are the most seriously impacted in the industrialised countries (other than the Netherlands) providing pork feed. Finally, the information function is primarily affected in the exporting developing countries. Knowing what impacts occur where is vital to developing more sustainable production systems not just taking into account sustainability in the Netherlands, but also in those countries providing feed for Dutch pigs. Since the latter may have ecosystems that are particularly vulnerable, it is of prime importance to localise every part of the production chain. The impact is not determined by the emissions (or resource use), on the one hand, or by the ecosystem, on the other hand, but by their combination only.

2.6 CONCLUSIONS[6]

This chapter has been devoted to the first hypothesis of the PROFETAS programme, that a shift from meat protein to plant protein is environmentally more sustainable than present trends. To this end, three approaches are presented to quantify the environmental performance of food chains. In short, we can conclude that a meat to plant protein shift indeed has a lot of potential for increasing the environmental sustainability of food production and consumption. Although the values provided differ among the economic, environmental and ecological approaches, there is a clear consensus both on which impacts are the most important, and on the NPFs chain scoring considerably better than the pork chain on sustainability, independent of the approach. Comparing the pork and pea-based NPFs chain more in detail, the three approaches found acidification could be reduced by a factor in the range 61-192. Additional reduction figures are in the range 6.4-111 for climate change, 6.0-14.3 for eutrophication, 3.3-47.5 for freshwater use and 2.8-9.5 for land use. Some differences are easily explained from the system boundaries. For example, from the values given in Figure 2-2, freshwater use reduction can be estimated as 62.3, which seems to be beyond the range given above. This is not the case, however, since about 20% of the freshwater use might be allocated to soy oil for human consumption, bringing the value below 50, and rounding brings it close to or below 47.5. Allocation differences are common to LCA-like estimates. The inclusion of

---

[6] Harry Aiking & Joop de Boer

links in the chain is also rather arbitrary, thus, transport is often disregarded, and since pea-NPFs are virtual products still, the processing and later stages were generally not taken into consideration. Other causes include the aggregation level and source of the data – either nationally or globally averaged.

Another important conclusion is that the economic, environmental and ecological approaches provide results that are clearly complementary. The outcomes of a conventional tool such as LCA are heavily dependent on the context in which the comparison is being made. The novel approaches developed by PROFETAS add context and more dynamic relationships, for example, as a result of the behaviour of agricultural producers and traders.

The economic activities leading to unsustainable emissions were the focus of Section 2.3. Linking economy and environment in a model has led to the important conclusion that for global environmental sustainability a local (European) transition in itself will be insufficient, because the centre of growing economies plus growing meat consumption is to be found in Asia and South America. Furthermore, it was clearly illustrated that the political (citizen) role of consumers who support more stringent environmental standards might be greater than their economic (consumer) role when they make trade-offs between the prices of pork and NPFs.

In Section 2.4 a new methodology was described, using ecosystem services as the theoretical basis underlying environmental sustainability. By no means fully mature yet, the method nevertheless supplies us with transparent criteria for a comparative study of alternative protein food options. In addition to its more solid theoretical foundation, it displays – as intended – several advantages over established tools such as LCA. Among the most prominent are the focus on effects, rather than on emissions, its low data requirements and its applicability to virtual products that are still to be produced. Presently, the method is limited to the agricultural production phase. After extension to the rest of the production and consumption chain, no doubt, the method can be generalised far beyond comparing protein food options.

Environmental indicators are addressed in Section 2.5, arguing that for developing a coherent *set* of indicators no framework turned out to be available, which (1) links pressures to environmental consequences, and (2) focuses on how each of the indicators can contribute to the problem solving logic required to answer questions of environmental sustainability. Rather, the existing frameworks take a one-dimensional perspective. Consequently, the process of indicator selection tends to be reduced to finding the best driving force, pressure, state, impact and response indicators for each individual environmental issue of concern. Therefore, it was argued that by working with causal networks, which highlight the multidimensional interconnections of processes, much more insight can be gained into the interrelation of indicators and, thus, how individual indicators can contribute to providing a more complete, holistic picture of what is happening in the environment. In short, process knowledge provides insight into cause-and-effect networks and can guide the selection of appropriate indicator sets.

Because a large part of the environmental impact of the pork chain takes place outside Western Europe, it is increasingly important to look at the global dimension of our production chains. Shifting to a more plant protein centred consumption pattern in Western Europe may reduce the environmental impact in the developing countries that presently meat and feed are imported from, but it may also have important economic repercussions in these countries. In turn, depending on what alternative livelihood strategies will be developed there, other negative environmental impacts may ensue. This goes to show that, even if our sole concern would be the physical environment, a reduction in the environmental impacts of food production cannot be achieved without taking into account global economic and social dimensions of changes in production chains. It also implies that for any changes in food production and consumption patterns to have a positive environmental effect, it is important to support food producers (farmers, traders and industry) in finding environmentally friendly alternative sources of livelihood.

Finally, there is yet another reason to look at the global picture. As was pointed out in Section 2.4, food production and consumption processes have a strong impact on a number of geochemical cycles vital to the ecology of this planet. Resource use, emissions and disturbance of natural processes take place locally, but through their effect on the carbon, nitrogen and hydrological cycles, their environmental impact has a global dimension.

REFERENCES

Aiking, H., and De Boer, J. (2004), "Food sustainability: Diverging interpretations", *British Food Journal*, Vol. 106 No. 5, pp. 359-365.

Baan, P.J., and Hopstaken, C.F.A.M. (1989), *Schade vermesting: Schade als gevolg van emissie van stikstof en fosfor en baten van emissiebeperking,* Ministerie van Volkshuisvesting, Ruimtelijke Ordening en Milieubeheer, Report 1990/7, The Hague.

Bolsius, E., and Frouws, J. (1996), "Agricultural intensification: Livestock farming in the Netherlands", in Sloep, P.B., and Blowers, A. (Eds.), *Environmental problems as conflicts of interest*, Arnold, London, pp. 11-36.

Bossel, H. (1996), "Deriving indicators of sustainable development", *Environmental Modeling and Assessment*, Vol. 1 No. 193, p. 218.

Brundtland, G.H. (1987), *Our common future*, Oxford University Press, Oxford, UK, World Commission on Environment and Development.

Carlsson-Kanyama, A. (1998), "Climate change and dietary choices: How can emissions of greenhouse gases from food production be reduced?", *Food Policy*, Vol. 23, pp. 277-293.

Carpenter, S.R., Caraco, N.F., Correll, D.L., Howarth, R.W., Sharpley, A.N., and Smith, V.H. (1998), "Nonpoint pollution of surface waters with phosphorus and nitrogen", *Ecological Applications*, Vol. 8 No. 3, pp. 559-568.

Costanza, R., Cleveland, C., and Perrings, C. (2000), "Ecosystem and economic theories in ecological economics", in Jorgenson, S.E., and Muller, F. (Eds.), *Handbook of ecosystem theories and management*, Lewis Publishers, Boca Raton, London, New York, Washington, DC, pp. 547-560.

Costanza, R., D'Arge, R., De Groot, R.S., Farber, S., Grasso, M., Hannon, B., Limburg, K., Naeem, S., O'Neill, R.V., Paruelo, J., Raskin, R.G., Sutton, P., and Van de Belt, M. (1997), "The value of the world's ecosystem services and natural capital", *Nature*, Vol. 387, pp. 253-260.

De Boer, J., Van der Grijp, N.M., Campins Eritja, M., and Gupta, J. (2004), "The conceptual framework", in Campins Eritja, M. (Ed.), *Sustainability labelling and certification*, Marcial Pons, Madrid, pp. 39-48.

De Groot, R.S. (1992), *Functions of nature: Evaluation of nature in environmental planning, management and decision making*, Wolters-Noordhoff, Groningen.

De Groot, R.S., Wilson, M.A., and Boumans, R.M.J. (2002), "A typology for the classification, description and valuation of ecosystem functions, goods and services", *Ecological Economics*, Vol. 41, pp. 393-408.

Erzberger, C., and Prein, G. (1997), "Triangulation: validity and empirically-based hypothesis construction", *Quality & Quantity*, Vol. 31, pp. 141-154.

FAO (2004), "AQUASTAT", Available at <www.fao.org/ag/agl/aglw/aquastat/main/index.stm>.

Fine, B., Heasman, M., and Wright, J. (1996), *Consumption in the age of affluence: The world of food*, Routledge, London.

Fonk, G., and Hamstra, A. (1995), *Future visions for consumers of Novel Protein Foods (Toekomstbeelden voor consumenten van Novel Protein Foods)*, DTO-Werkdocument VN12, Duurzame Technologische Ontwikkeling, Delft, The Netherlands.

Ginsburgh, V.A., and Keyzer, M.A. (1997), *The structure of Applied General Equilibrium models*, MIT Press, Cambridge (Mass.).

Goosen, H., Janssen, R., Verhoeven, M.L., Omtzigt, N., and Verhoeven, J.T.A. (2004), *Ruimtelijke multi-criteria analyse in veenweidegebieden*, Institute for Environmental Studies, Vrije Universiteit, Amsterdam.

Hammond, A., Adriaanse, A., Rodenburg, E., Bryant, D., and Woodward, R. (1995), *Environmental indicators: A systematic approach to measuring and reporting on environmental policy performance in the context of sustainable development*, World Resources Institute, Washington, DC.

Helms, M. (2004), "Food sustainability, food security and the environment", *British Food Journal*, Vol. 106 No. 5, pp. 380-387.

Helms, M., and Aiking, H. (2003), "Food and the environment: Towards sustainability indicators for protein production", in Tiezzi, E., Brebbia, C.A., and Usó, J.L. (Eds.), *Ecosystems and sustainable development Vol. 2*, WIT Press, Ashurst (UK), pp. 1047-1056.

Jackson, L.E., Kurtz, J.C., and Fisher, W.S. (2000), *Evaluation guidelines for ecological indicators*, EPA/620/R-99/005, Environmental Protection Agency, Washington, DC.

Keyzer, M.A., Merbis, M.D., Pavel, I.F.P.W., and Van Wesenbeeck, C.F.A. (2003), *Diet shift towards meat and the effects on cereal use: Can we feed the animals in 2030?*, Available at <www.sow.econ.vu.nl/pdf/wp01-05R.pdf>, Centre for World Food Studies, Vrije Universiteit, Amsterdam.

Niemeijer, D. (2001), "Assessing environmental sustainability: Process-based models versus state indicators", Proceedings 97th Annual Meeting of the Association of American Geographers, New York, pp. 1-4.

Niemeijer, D. (2002), "Developing indicators for environmental policy: Data-driven and theory-driven approaches examined by example", *Environmental Science & Policy*, Vol. 5 No. 2, pp. 91-103.

Niemeijer, D., and De Groot, R.S. (2005a), "A conceptual framework for selecting environmental indicator sets", *(in preparation)*.

Niemeijer, D., and De Groot, R.S. (2005b), "Framing environmental indicators: Moving from causal chains to causal networks", *(in preparation)*.

Niemeijer, D., Helms, M., and Zhu, X. (2005), "Environmental pressure indicators for food production: Comparing plant and meat protein production chains", *(in preparation)*.

NRC (2000), *Ecological indicators for the nation*, National Academy Press, Washington, DC.

OECD (1993), *OECD Core set of indicators for environmental performance reviews: A synthesis report by the Group on the State of the Environment*, Environment Monographs Vol. 83, Organisation for Economic Co-operation and Development, Paris.

OECD (1999), *Environmental indicators for agriculture: 1. Concepts and frameworks*, Organisation for Economic Co-operation and Development, Paris.

Shoven, J.B., and Whalley, J. (1992), *Applying General Equilibrium*, Cambridge University Press, Cambridge.

Smeets, E., and Weterings, R. (1999), *Environmental indicators: Typology and overview*, Available at <reports.eea.eu.int:80/TEC25/en/tech_25_text.pdf>, European Environment Agency, Copenhagen, Denmark.

Smil, V. (2001), *Enriching the earth: Fritz Haber, Carl Bosch, and the transformation of world food production*, MIT Press, Cambridge (Mass.).

Smil, V. (2002), *The Earth's biosphere: Evolution, dynamics, and change*, MIT Press, Cambridge (Mass.).

Van Doorn, J. (1993), *Energie uit Nederlandse biomassa*, Bijdrage aan de 4e Nederlandse zonne-energie conferentie, 1-2 april 1993 te Veldhoven, Energie Onderzoek Centrum Nederland, Petten, The Netherlands.

WRI (2004), "Earth Trends - The environmental information portal", Available at <earthtrends.wri.org>, World Resources Institute, Washington, DC.

Zhu, X., and Van Ierland, E.C. (2003a), "Modeling consumers' preferences for novel protein foods and the environmental quality: An AGE approach", Proceedings 12th EAERE Conference, Bilbao, Spain.

Zhu, X., and Van Ierland, E.C. (2003b), "Protein chains and environmental pressures: A comparison of pork and Novel Protein Foods", *Environmental Sciences*, Vol. 1 No. 3, pp. 254-276.

Zhu, X., Van Ierland, E.C., and Wesseler, J. (2004), "Introducing novel protein foods in the European Union: Economic and environmental impacts", in Evenson, R.E., and Santaniello, V. (Eds.), *Consumer acceptance of genetically modified food*, CAB International, Oxfordshire, UK, pp. 189-208.

CHAPTER 3

TECHNOLOGICAL FEASIBILITY

Johan M. Vereijken, Lynn Heng, Francesca E. O'Kane,
Xinyou Yin, Jan Vos, Egbert A. Lantinga,
Emmanouil N. Tzitzikas, Jean-Paul Vincken,
Krit Raemakers, Richard G.F. Visser,
Radhika K. Apaiah, Frank Willemsen & Martinus A.J.S. van Boekel

3.1 INTRODUCTION[1]

3.1.1 Background

The PROFETAS programme tries to verify three hypotheses (see Chapter 1). The common aim of the projects described in this chapter is to assess the second hypothesis, regarding the technological feasibility of producing NPFs based on pea proteins. The starting point was the desk study on NPFs, which had been performed within the framework of the Dutch programme on Sustainable Technology Development (see also Chapter 1). The main goal of the latter study was to identify ways to reduce the environmental impact of food systems by a factor of about 10 to 20 by 2035 as compared to 1995. It was concluded that a replacement of 40% of meat by NPFs could, depending on the source of the protein, result in such a reduction (Weaver *et al.*, 2000). In particular, products based on proteins from micro-organisms such as *Spirulina* (a cyanobacterium fixing its own CO_2) appeared to be very promising. Despite this promising potential, it was decided to focus the PROFETAS programme on plant-derived NPFs because:

- According to the study mentioned above NPFs based on plant proteins are very attractive with respect to the reduction in environmental impact that can be achieved. Plants fix their own CO_2, just like *Spirulina,* but in contrast to most other organisms (including most micro-organisms), thus avoiding the resource-inefficient conversion of carbohydrates.
- Proteins derived from plants are expected to remain cheaper than those derived from micro-organisms, mainly because of the costs involved in the cultivation step of the latter.

[1] Johan M. Vereijken

H. Aiking et al. (eds.),
Sustainable Protein Production and Consumption: Pigs or Peas?, 51–98.
© 2006 *Springer. Printed in the Netherlands.*

- Compared to organisms like *Spirulina*, much more is known about the primary production of plants and about their proteins (including processing).

NPFs in the form of meat substitutes are already on the market. Therefore, these products will be touched upon below, particularly the source proteins and processes used to produce them (Section 3.1.2). As was already mentioned in Chapter 1 (and to be more thoroughly discussed in Section 6.2), the pea was selected as the source from which the proteins are derived to produce NPFs. Therefore the pea and pea proteins, in particular their composition and industrial extraction, will be described briefly in Section 3.1.3.

To assess the technological feasibility of producing NPFs based on pea proteins a number of projects were carried out. The coherence among these technological projects is described in the final part of this introductory section.

In subsequent sections of this chapter the technological projects are described in more detail, with an emphasis on their results, conclusions and implications for NPFs production. In the concluding section of this chapter, the technological feasibility of producing NPFs will be discussed.

3.1.2 Currently available meat substitutes

Source proteins

The introduction of meat substitutes (also termed meat alternatives, meat analogues or meat replacers) on the Western markets is a relatively recent development, starting in the beginning of the 1960s (Sadler, 2004; Davies and Lightowler, 1998). However, what we call meat substitutes includes protein-rich products such as tofu and tempeh which have been eaten in Asia for centuries. Both products are based on soy protein. Tempeh is produced by fermentation of soybeans and tofu is more or less a kind of cheese. It is the pressed curd of soy proteins coagulated by the addition of calcium sulphate or calcium chloride.

The introduction of meat substitutes in addition to these traditional Asian products started with the production of Texturised Vegetable Protein or TVP (also known as Texturised Soy Protein or TSP). This "spongy" product is produced by extrusion (see below) of defatted soy meal or of soy concentrates (i.e. defatted meal from which soluble carbohydrates have been removed by a washing step). TVP is available in the form of granules and chunks. Apart from being used as a starting material for meat substitutes, TVP is also used for incorporation into meat products as a so-called "meat extender", serving both as a relatively cheap substitute for meat proteins and to improve products properties.

Since the 1960s many meat substitutes have been launched that are based on TVP or comparable soy protein products. In more recent years proteins other than those from soy have been introduced as the starting materials for the production of meat substitutes. Especially wheat gluten, the protein fraction remaining from wheat flour after the removal of starch, has to be mentioned. In about 1992, the company Kerry Ingredients launched Wheatpro™, an extruded form of wheat gluten. Furthermore, extruded combinations of wheat gluten and other proteins (soy but also pea proteins) are on the market. These combinations are designed to better mimic the fibrous texture of meat muscle than the more "spongy" TVP. Very recently, products such as burgers based on lupine proteins have been launched.

Apart from plant proteins, meat substitutes based on a fungal protein are available. They are more or less the "spin-off" of a large development that started in the 1960s to produce single cell protein from fungi that originally reside in the soil. The fungus *Fusarium venenatum*, also called *Fusarium graminearum*, is produced by fermentation in stirred tanks. After harvesting, the nucleic acid content is reduced and the biomass (called mycoprotein) is mixed with egg albumin as a binding protein to produce Quorn™.

From the above it can be concluded that only a limited number of plant proteins are used to produce meat substitutes, the major ones being soy proteins and to a lesser extent wheat gluten.

Processes

As has been pointed out above, the main process to produce the starting material for meat substitutes is extrusion. Extruders (Figure 3-1) consist of a barrel in which the rotation of one or more screws transports the material from the feed site to the exit. Due to this rotation a pressure is built up that is needed to push the material through the die at the exit, where the material is shaped. The mechanical energy input of the rotating screw(s) is partly converted to heat by internal friction. Furthermore, the barrel is equipped with heating and cooling jackets to provide an adjustable temperature profile along the extruder axis. While the material is transported from the feed site to the exit, it is subjected to various unit operations, such as mixing, homogenizing, compressing, shearing, elongating, and heating/cooling. In the extruder the proteins are subjected to shear and heat, resulting in their unfolding and aggregation and hence "texturisation". The resulting shape is determined by the configuration of the die.

After preparation of the base material, meat substitutes are produced by conventional processes used in the meat industry. Depending on the type of base material and the products aimed for, these processes may comprise cuttering (reduction in size), mixing (with other ingredients), forming, seasoning, battering, and slicing/dicing.

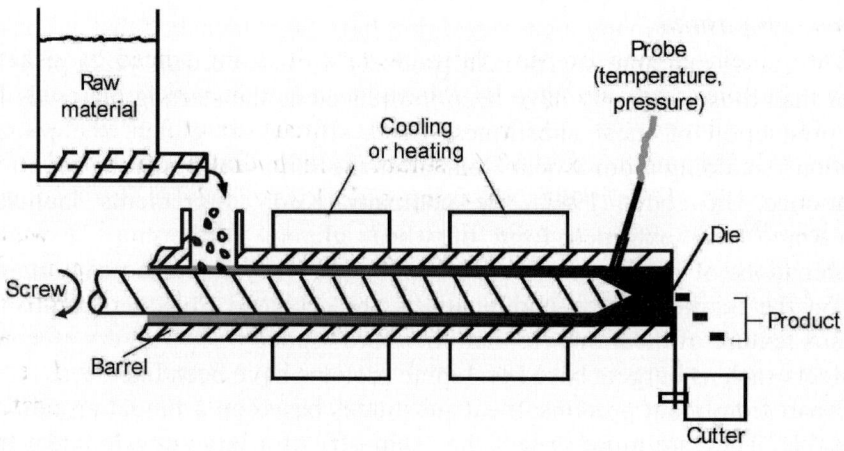

Figure 3-1. Schematic drawing of an extruder.

3.1.3 The pea and the pea proteins

Peas (*Pisum sativum* L.) belong to the *Leguminacea* (or *Papilionacea*), a family of which most members are capable of fixing nitrogen from the air due to their symbiosis with nitrogen fixing bacteria. For this reason they need no or only small amounts of added nitrogen fertilisers during their cultivation. Other members of this very large family are lentils (*Lens culinaris*), all kinds of beans (such as French beans, *Phaseolus vulgaris,* and faba beans, *Vicia faba*), chick peas (*Cicer arietinum*) and groundnuts (*Arachis hypogaea*). From an economic point of view the most important member of this family is soy (*Glycine max*). With respect to cultivated area, world production and international trade, soy is far more important than the other leguminous species.

Most of the *Leguminacea* contain in their seeds either high amounts of oil and low amounts of starch (e.g. soy) or the reverse, high amounts of starch and low amounts of oil (e.g. peas) as an energy reservoir for the growing seedling. Compared to crops such as cereals, they all contain relatively high amounts of protein and are therefore called protein crops.

Peas are used for both food and feed applications. For food applications they are harvested in an immature state. Then they taste sweet because not all sugars have been converted to starch. They are consumed as green peas but most of them are preserved by canning or freezing and sold as such. For feed applications, quantitatively by far the most important outlet, peas are harvested in the mature and dry state.

Protein composition

The protein content of peas varies widely. It is influenced by genetic factors, but also very strongly by the environment in which the peas are grown (soil, climate, etc.). On average, peas contain about 25% protein, the remainder being starch (about 50%), soluble carbohydrates, fibres, and some minerals and fat. The proteins are composed of water soluble and insoluble proteins, the latter accounting for 10 to 25% of the total protein. The water-soluble proteins are divided into albumins and globulins, accounting for about 15 to 25% and 55 to 65% of the total protein content, respectively (Owusu-Ansah and McCurdy, 1991; Gueguen, 2000). The albumins are proteins with a physiological function in the seed (enzymes, plant defence proteins etc.). The globulins serve as storage proteins, a reservoir of protein for the growing seedling. Globulins, being the major protein fraction in peas, are composed of two major fractions (called legumin and vicilin) and a minor one called convicilin. The ratio between the globulins is very variable, depending on genetic and environmental factors.

Industrial pea protein extraction

On an industrial scale protein preparations from the pea may be obtained by two different processes: a wet and a dry one. In the dry process, peas are milled to small particles. Such grinding results in flours containing populations of particles differing in size, shape and density. Subsequently the particles are separated from each other in a stream of air and fractionated into a coarse fraction (starch enriched) and a fine fraction (protein enriched). The practical limit to which pea proteins can be purified by this technique – called air classification – appears to be about 65%; protein preparations having such protein content are termed concentrates. Total protein recovery by the dry procedure is on the order of 60-70% (Gueguen, 2000), because some of the protein is still present in the coarse fraction.

In the wet process, the proteins are extracted at alkaline pH. During this extraction also other components are solubilized, particularly some carbohydrates. The removal of these solubilized carbohydrates can either be performed by ultrafiltration or by iso-electric precipitation. In the latter procedure the pH of the extract is lowered to such a value that the majority of the proteins precipitate (the iso-electric pH of the protein mixture). After centrifugation the precipitate is dried. Most of the albumins (and some of the globulins) remain soluble, resulting in a loss of protein. By ultrafiltration the extract is purified by passing it through membranes that selectively remove small molecules such as the solubilized carbohydrates and some small albumins. Wet processing usually results in protein preparations with a protein content of about 90%; such preparations are termed isolates. Because of the loss of the insoluble proteins and part of the albumins (the latter especially during iso-electric precipitation) the globulins are by far the most abundant proteins in pea isolates. Total protein recovery of the two wet processing procedures differs: iso-electric precipitation has a protein yield of about 70% (Gueguen, 1983; Gueguen, 2000) and the ultrafiltration procedure has a higher yield (up to 85%, Vose, 1980; Gueguen, 2000).

3.1.4 Technological projects

The market for meat substitutes is still quite small. In the Netherlands, the share of meat substitutes is only about 1% of the total market for meat and meat products. This is likely to be at least partly due to the fact that present meat substitutes do not meet the consumer preferences with respect to sensory quality (Van der Lans, 2001). Especially bite, taste, and juiciness score low in comparison to meat.

To obtain a larger market share, NPFs should be developed that align better with consumer demands than present meat substitutes. To this end, the consumer should be placed central in the development of NPFs, a so-called consumer oriented approach. As a start, an inventory should be made with respect to consumer behaviour, demands and preferences within the diverse target groups. As a next step, these preferences should be translated into physical parameters, which can be used to direct the processing of NPFs. In turn, this processing may result in requirements for the starting material for NPF production, i.e. the crops from which the proteins are derived. The quality of this material can be affected both by cultivation and by breeding.

To cover this whole chain of consumer-oriented NPF development, seven projects were formulated in the PROFETAS programme (Figure 3-2). Two projects are focussed on the consumers, their behaviour, demands and preferences and the translation of these preferences to physical product parameters. These projects will be dealt with in Chapter 4. Two projects are focussed on the processing of NPFs and more particularly on main shortcomings of present meat substitutes: their flavour (including taste) and texture (determining "bite"). These are the subjects of Sections 3.2 and 3.3, respectively. Two projects are devoted to primary production, i.e. one to cultivation (Section 3.4) and one to breeding (Section 3.5). One integrative project is focussed on optimising the whole chain of NPF production, from consumption to breeding (Section 3.6).

It should be emphasised that the intention of the PROFETAS programme is to develop tools and know-how to produce NPFs. The actual production of NPFs is the domain of manufacturers. On the one hand, based on market and marketing studies and, on the other hand, on economic and strategic considerations, the manufacturers should decide what products will be made for which groups of consumers. This decision determines the textures and flavours that are required.

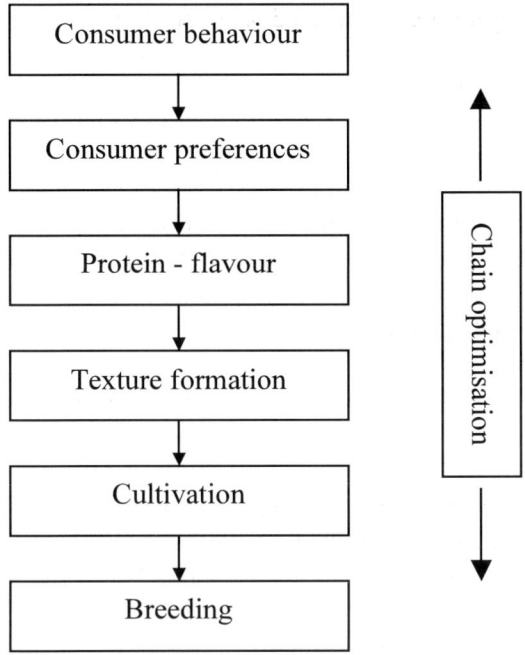

Figure 3-2. Consumer-oriented NPF development.

Apart from this set of projects, directed at producing NPFs, another technological project was started (Section 3.7), aimed at providing an inventory of possible applications of the non-protein fractions of crops being used for NPF production. In the case of peas, these non-protein fractions are mainly starch and polysaccharides derived from cell walls, but for other crops oils may be produced instead of starch. The non-protein fractions will become available on quite a large scale if 40% of meat consumption is to be replaced by NPFs. For the economic feasibility of producing proteins, the value of the applications of the non-protein fractions may be of decisive importance: soy processing, which has taken enormous proportions, rests on both the values of its protein and its oil.

As stated in the beginning of this introduction, the overarching aim of the technological projects is to assess the technological feasibility of producing NPFs so well that a transition from meat proteins to plant proteins may occur. In the concluding section of this chapter (3.8), the outcomes of the individual projects are discussed in the context of this overarching aim.

3.2 PROTEIN–FLAVOUR INTERACTIONS[2]

3.2.1 Introduction and research goal

Consumers will rate NPFs according to the quality perception during consumption. An important aspect in quality perception is flavour. Flavour encompasses several sensory characteristics such as taste, smell and odour, which are sensations brought about by molecules that activate receptors in the mouth and nose. Flavour molecules may be naturally present in foods, or may arise during food preparation (e.g. cooking). They may also be added as flavourings. Flavours generally perceived as undesirable by a given target group are termed "off-flavours".

In contrast to foods such as fruits, vegetables and meat, NPFs, being fabricated foods, do not contain natural flavours. Proteins extracted from vegetable sources such as peas can form the basis for NPFs. With the exception of a very few, proteins themselves do not possess any taste. However, undesired flavour molecules may accompany proteins. Notorious are for instance the volatile flavour compounds that are present in many soy protein preparations, giving rise to the typical beany flavour that is generally not appreciated by Europeans. The flavour perceived from a protein-based product, whether desirable or undesirable, is due to the adsorption of flavour molecules to the proteins, which are known to interact strongly with small, volatile molecules such as most flavour compounds. Hence, there are two main questions with regard to the proteins used for NPFs production. One concerns the presence of undesirable, volatile flavour compounds in the proteins. The other one deals with the strength of the adsorption of desirable, volatile flavour molecules to the proteins, because these molecules have to be released during consumption in order to be perceived by the consumer. Thus, the research goals of this project were to investigate the presence of volatile flavour compounds that are present in pea proteins, and how strongly pea proteins adsorb volatile flavour compounds as a function of processing conditions. The influence of flavour compounds on protein structures was also investigated, because structural changes may affect other properties of pea proteins relevant for NPF production. In addition, a non-volatile class of flavour compounds, the saponins, which are known for their bitterness, were investigated in different pea varieties.

[2] Lynn Heng

The knowledge obtained from this research will enable meeting consumer preferences with respect to flavour by adding flavour compounds during production of NPFs or by masking "off-flavours". Furthermore, this knowledge will enable breeders to select or breed suitable pea varieties.

3.2.2 Methods and results

In this study, the two major proteins from the pea, legumin and vicilin, were used. These are salt-extractable globular proteins, which account for 65-80% of the soluble proteins present in peas (Owusu-Ansah and McCurdy, 1991). In this project, pea proteins were extracted at pH 8, followed by a preliminary batch-wise purification. Purified legumin and vicilin were subsequently obtained by anion-exchange chromatography.

The purified proteins were used for protein-flavour interaction studies with volatile flavour compounds. To facilitate experimental design and interpretation of results, model flavour compounds were used as representatives of complex mixtures that are usually applied by flavourists. The model compounds used for the studies were simple aldehydes and ketones. Static headspace-gas chromatography was used to study protein-flavour binding. This technique allows the detection of volatile flavour compounds in the headspace above a protein solution. The difference in the concentration of flavour compound between a sample containing only volatile flavour in solution, and a sample containing volatile flavour and protein, is the number of flavour molecules adsorbed by the protein. To assess the effects of processing conditions on protein-flavour interactions, static headspace-gas chromatography was performed at different protein and flavour concentrations, as well as at various conditions of pH and temperature. Many interaction studies have been done to show binding of flavour molecules to proteins such as soy proteins (Chung and Villota, 1989; O'Keefe et al., 1991a; O'Keefe et al., 1991b) and milk proteins (Andriot et al., 1999; Relkin et al., 2001). However, only a few were performed on pea proteins (Dumont, 1985; Dumont and Land, 1986).

The present study showed that the two protein preparations did contain volatile flavour compounds, most likely originating from fat oxidation. Furthermore, it was shown that vicilin binds both aldehydes and ketones, but legumin only showed binding to aldehydes under the conditions used (Heng et al., 2004). The presence of flavour molecules, however, did not have an influence on the molecular structures of the proteins themselves. The amount of binding increased with flavour concentrations. Legumin showed binding to aldehydes only at pH 7.6 and not at pH 3.8. Heating of vicilin was observed to lead to a decrease in the binding of aldehydes and ketones to the protein (Heng *et al.*, 2004). This is an important consideration for food processing, in which heating is the most common treatment.

Peas are also known to contain non-volatile flavour components called saponins (Bishnoi and Khetarpaul, 1994; Daveby *et al.*, 1997; Daveby *et al.*, 1998). These compounds are known to contribute to the bitter and metallic taste perceived in stored pea flour (Price and Fenwick, 1984; Price *et al.*, 1985) and are, therefore, important in flavour studies. In addition, saponins were also found to interact with proteins (Ikedo *et al.*, 1996; Potter *et al.*, 1993; Shimoyamada *et al.*, 1998; Shimoyamada *et al.*, 2000) and may thus affect the binding of other flavour compounds to the proteins. Hence, it is relevant to know the types and quantities of saponins present in peas. With the use of high performance liquid chromatography coupled with mass spectrometry (HPLC-MS), two main groups of saponins were identified in pea flour; the DDMP saponin and saponin B. DDMP saponins are very heat labile and are easily converted to group B saponins at elevated temperatures (Kudou *et al.*, 1992; Heng *et al.*, 2005).

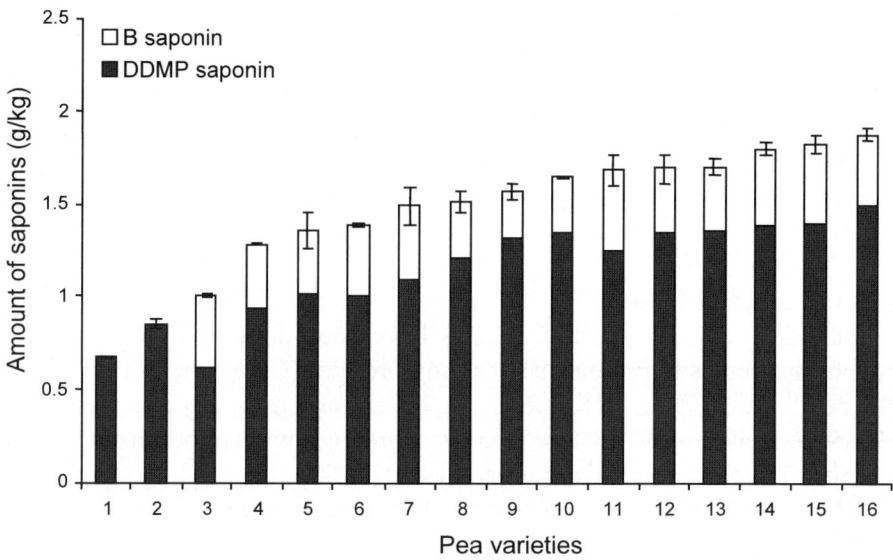

Figure 3-3. DDMP and B saponins content of 16 different pea varieties (1: KPMR 146 India; 2: Pisum elatius, 1140175 (CGN 10205); 3: Cisca; 4: FAL 49110 Mongolie; 5: CEB 1475.1; 6: Baccara; 7: CEB 1466; 8: Pisum arvense (CGN 10193); 9: Pisum elatius, Marbre (CGN 03351); 10: Courier; 11: 6 S 41.4; 12: Solara; 13: NGB 102149 Iran; 14: Solido; 15: Supra; 16: NGB 101293 Jordaan).

Pea varieties were found to differ significantly in their saponin concentrations (Figure 3-3). DDMP saponin was found to be the predominating saponin in all the varieties. The total saponin contents between 16 different pea varieties varied from 0.7-1.9 g/kg. Sensory evaluation on the bitterness of DDMP saponin and saponin B showed a correlation between the saponin concentration in peas and the perceived bitterness. DDMP saponin was shown to be more bitter than saponin B. The stability of DDMP saponin was also studied. DDMP saponin was found to be converted easily to saponin B at temperatures over 40 °C, and has an optimum stability at about pH 7 (Heng et al., 2005). These findings offer food technologists the possibility to convert the more bitter DDMP saponin to the less bitter saponin B.

3.2.3 Concluding remarks

This study has provided details about the interactions of the two major pea proteins, vicilin and legumin, with aldehydes and ketones. Legumin was found to bind less of these flavour compounds than vicilin under the same conditions. Furthermore, the amount of flavour compounds associated with the proteins – under various conditions of pH and heat treatment – can be estimated from the results obtained. This enables flavourists to meet consumer preferences with respect to taste.

It has also been shown that pea varieties differ in saponin content, both with respect to amount and type of saponins. The bitterness in peas due to the presence of saponins is of importance for food manufacturers. As an alternative to selecting pea varieties that are low in saponin content, pea processors can explore ways to remove these compounds from pea protein preparations. Furthermore, food manufacturers can reduce bitterness by applying a heating step because, as has been shown in this investigation, heating will convert the more bitter DDMP saponin to the less bitter saponin B.

Despite the information provided by this study, there are still aspects with respect to the flavour of NPFs that require further investigation. Two important aspects are:
- In addition to proteins, other components present in NPFs can also influence flavour perception. Many volatile flavour components are soluble in fats and, hence, the presence of fats will affect the release of flavour volatiles during food consumption. Therefore, the behaviour of flavour components in a more complex food system and, ultimately, in a food product should also be studied. Subsequently, sensory evaluation by trained or consumer panels has to be performed. The type of flavour to be applied to a product is dependent on the type of product and the target

consumer segment. This is the phase of product development, which is part of a company's production and marketing strategy.
- Apart from saponins also other compounds present in pea proteins may result in an undesirable flavour. Important in this respect is the finding that volatile flavour compounds, probably originating from fat oxidation during isolation of the proteins, are present in the pea proteins used in this study. To ease the flavouring of the end products, manufacturers usually prefer as starting material to have protein preparations that are free of such compounds. To produce these preparations knowledge is required about the prevention of the formation of volatile flavour compounds during protein preparation or/and their removal.

3.3   NPF TEXTURE FORMATION[3]

3.3.1   Introduction and research goal

The acceptance of NPFs by consumers depends on many factors. One of the product-related factors is texture, which is not only important for the appearance of the product but also for the sensations the consumer experiences when eating the product. The mouth feel and the resistance to fracture during chewing are very important characteristics with respect to final product evaluation. Texture is a quality cue, meaning that a consumer already has an opinion on the quality of a product just by looking at it. The first impression of a product, whether it has a solid, weak or deformable appearance has a large impact.

This project was not about what texture an NPF should have, but about the ability of giving a certain texture to a product by studying texturising properties of the proteins involved. Texture as such is not a clear concept. Textures vary widely; they can be fibre-like (such as in meat products), gel-like (such as in yoghurt), coagulated (such as in cheese, tofu), etc. Since a particular desired texture for an NPF is not yet known, it was decided to study the heat-induced gelling behaviour of pea proteins. The reasoning behind this was that almost certainly NPFs will be heated, in the factory as (part of) the process to obtain a certain texture and/or by the consumer when preparing NPFs at home. So knowledge about the heat-induced gelling behaviour of pea proteins will be relevant, no matter what textures

[3] Francesca E. O'Kane

eventually will be needed. Furthermore, it was hypothesized that knowledge of heat-induced gelling behaviour of the individual major pea protein fractions is of utmost importance, if not a prerequisite, for understanding and directing texture formation of industrially prepared pea protein preparations. Therefore, the research question was how individual pea protein fractions would behave upon heating and how the ensuing structure formation comes about.

3.3.2 Methods and results

Pea protein fractionation and characterisation

An isolation and purification procedure was used throughout this study to obtain purified storage proteins for further investigations. Emphasis was on storage proteins, because they comprise the majority of the proteins present in pea seeds and dominate the functionality of pea protein preparations such as isolates. It was known from the literature that peas contain two major storage proteins, i.e. vicilin and legumin, and a minor one called convicilin. An unexpected separation of vicilin into two fractions differing in amount of convicilin triggered additional work on the characterisation of pea vicilin (O'Kane, 2004; O'Kane *et al.*, 2004a). To this end, a series of chromatographic techniques were used to remove the convicilin that heavily contaminated one of the vicilin fractions. This separation was not possible, and instead the results obtained indicated that convicilin is not a contaminant, but a subunit of the extracted vicilin protein. This was further supported by various techniques (gel electrophoresis, differential scanning calorimetry, circular dichroism and solubility experiments). Convicilin was denoted as the α-subunit of vicilin because of the resemblance (such as a highly charged N-terminal part) to corresponding proteins (the α and α' subunit of β-conglycinin) present in soy. In conclusion, this part of the study clearly showed that vicilin is more heterogeneous with respect to subunit composition than expected based on literature data, because convicilin, now termed α-subunit, is one of the subunits.

Gelling behaviour of individual protein fractions

The implications on gelling behaviour of the vicilin subunit heterogeneity was explored by comparing the two vicilin fractions described above (with differing subunit compositions), under the same conditions, with regard to gelling properties upon heating and cooling (O'Kane, 2004; O'Kane *et al.*, 2004b). Rheological techniques were used as the main instrument; these techniques basically measure resistance to deformation of structures and provide information about their firmness and elasticity. As is usual for heat-induced protein gelation, a three-step process was observed, first denaturation (unfolding) and then aggregation of denatured proteins, followed by further aggregation into network structures upon cooling. It was clearly demonstrated that at near-neutral pH the highly charged N-terminal part of the α-subunits contributed a net negative charge that ultimately hindered extensive gel network formation, i.e. increased the minimum gelling concentration. An interesting observation was that the α-subunits caused the formation of transparent gels. Transparency is a characteristic of gels that indicates that the aggregates that make up the gel network are small in size. This observation is in keeping with proteins that carry a large net negative charge. The finding was further substantiated by the observation that the inhibiting effect of the α-subunits disappeared at low pH as well as at high ionic strength. Another observation was that mixing of the vicilin fraction containing a high amount of α-subunits with legumin had an effect on the transparency of the resulting gel. It was concluded that the subunit composition of vicilin can be targeted as a tool to modify the gelation of these pea proteins.

Having explored vicilin, a series of experiments was done to compare the intermolecular forces that create and support different pea legumin gels (O'Kane, 2004; O'Kane et al., 2004c). The techniques used were, among others, differential scanning calorimetry (giving information about, among others, the thermal stability of proteins) and rheology. In view of the importance of sulphydryl groups, use was made in some experiments of N-ethylmaleimide (NEM), a compound that blocks sulphydryl groups. Overall, it was shown that pea legumin gel formation was not effected by changes in the heating rate. However, changing the cooling rate affected gel strength: a reduced cooling rate resulted in stronger gels, and hence more and/or stronger interactions between the proteins. The addition of NEM reduced this effect, indicating that disulfide bond formation is involved in strengthening during cooling. The addition of NEM did not affect gel strength during the heating phase, so disulfide bonds are not essential for the formation of the initial gel network.

Comparison of legumin with the molecularly homologous soy glycinin gelled under the same conditions enabled the hypothesis to be tested that a common model for legumin-like protein gelation can be built based on molecular reasoning. The denaturation temperatures as determined by differential scanning calorimetry were found to be very similar. Furthermore, the two proteins responded in a similar way to changes in conditions used, indicating that the same physical and chemical driving forces were acting during gelation. However, the minimal gelling concentration between the proteins differed, that of legumin being higher than that of glycinin. More strikingly, when applying a procedure for reheating and recooling the two proteins responded differently: in contrast to glycinin gels, which were not affected by this procedure, legumin gels became stronger. So no common model could be built for legumin-like protein gelation.

Gelling behaviour of protein mixtures

A comparison of the gelation behaviour of protein isolates from five different pea cultivars was done (O'Kane, 2004; O'Kane *et al.*, 2005). The cultivars differed in legumin and vicilin content. The isolates were prepared by iso-electric precipitation and were virtually free of albumins. This study also enabled us to explore to what extent the gelation behaviour could be predicted based on knowledge of how the individual storage proteins gelled. One cultivar (Solara) stood out for its ability to form strong gels both in the presence and absence of NEM, indicating that the gel formed independently of any disulfide bond formed by legumin. In contrast, the other cultivars formed stronger gels only when NEM was added and so disulfide bond formation was involved. The purified legumin from cultivar Solara also exhibited gel-strengthening ability when added to a protein isolate from another cultivar. The addition of purified legumin from cultivar Supra had nearly no effect on gel strength, indicating that the gel-strengthening ability of legumin is cultivar specific. Determination of the free-thiol and total cysteine contents of purified legumin and protein isolates highlighted no direct relation between the disulfide bonding potential of the cultivars and their gelling behaviours, so an indirect explanation for why NEM affected the gel strength for all pea cultivars except Solara was put forward. This was that the spatial arrangement of SH-groups has an effect on the types of aggregates formed and consequently on the resulting network formation. Isolates of two cultivars (Finale and Espace) gave only weak gels and this could be related to the content of α-subunits of vicilin in these isolates, in line with the findings reported above. Overall, it was concluded that the vicilin α-subunit influence on gelation behaviour is a quantity dependent behaviour. That of legumin is variety specific, however, and depends on protein characteristics as yet unidentified.

3.3.3 Concluding remarks

Much information has been obtained both on pea protein composition and on gelling behaviour of pea proteins. With respect to the former, it has been shown that convicilin should not be regarded as a separate pea protein fraction but as a subunit of vicilin. Analogous to the nomenclature of soy proteins, we proposed to denote convicilin as the α-subunit of vicilin. With respect to gelling behaviour, an important finding was that this subunit has a negative effect on the gelling behaviour of pea proteins at near neutral pH, a pH most likely very near to the one of NPFs. In order to improve the performance of pea protein for NPF production, a reduction in the amount of this subunit, by for example breeding, is likely to be an interesting target.

Another important finding is that gelling behaviour of protein mixtures, analogous to ones commercially produced, is affected by the cultivar from which the proteins are isolated. This offers both opportunities (preparation of a range of gels differing in properties by choosing the right cultivar) as – perhaps – difficulties (dependence on a small range of cultivars) for producing NPFs. Comparison with the analogous soy proteins showed next to similarities (e.g. the effect of subunit composition of pea vicilin or soy β-conglycinin on gelling behaviour) differences that cannot be explained on the basis of present knowledge. So plant proteins can be different in terms of gelling behaviour, even though they look quite similar on the molecular level.

The present research project has led to considerable insight into the behaviour of legume proteins, in particular pea proteins, with respect to heating and subsequent aggregation and gelling. This knowledge can be used to go to a next phase in the understanding of texturising properties of proteins. Again it should be emphasized that it is not directly the task of a programme like PROFETAS to design a specific texture. Rather, the task is to supply knowledge and various options to reach certain textures, in other words to provide a guideline for product developers in a company.

Which textures ultimately are designed depends on marketing research to find out what textures are appealing to the different consumer groups. This will most likely result in a broad range of appealing textures. So, it remains of utmost importance to understand texture formation, because this understanding enables control and prediction of texture formation in NPFs, tools indispensable for directed product development. There are two major challenges. The first is how to obtain various types of texture using pea or other plant proteins. This requires a more process-technological approach, using techniques such as extrusion and spinning or novel technical options such as texturising by phase separation. The second challenge is to combine proteins with other food components, such as starch or other polysaccharides, and fats. After all, NPFs will not just consist of proteins. The gelling behaviour of such mixtures may be different from relatively pure proteins as used in the present study.

3.4 DESIGNING SUSTAINABLE PLANT-PROTEIN PRODUCTION SYSTEMS[4]

3.4.1 Introduction and research goal

The NPFs within the PROFETAS programme are produced from proteins that are derived from plants. Therefore, an indispensable component of the PROFETAS programme is to gain insight into the options for sustainable production of the raw plant protein material. This project is directed towards gaining this insight and aims to design tools and strategies for the development of sustainable systems for the primary production of plant protein.

In the PROFETAS programme, the pea was chosen as the model crop from which proteins are derived to produce NPFs. The pea was selected primarily because of its protein content, its ability to grow in Western Europe, and the absence of unwanted substances (see also 6.2). Seed yield per hectare is dependent both on the cultivar (or more correctly, the genotype) and the environment (soil, temperature, water availability, day length etc.). It was the objective of this project to design a model for the response of pea genotypes to different environments. By using such a model an optimal pea production system can be defined with respect to quantity and quality of pea seeds and to resource use efficiency. Furthermore, potential pea producing areas can be identified.

3.4.2 Methods and results

It was a major task to develop a robust model for predicting crop yield and protein production in response to genotypic characteristics and environmental variables. The model developed was subsequently subjected to evaluation analysis. The evaluated model was then applied to map seed yield and seed protein production in Europe for three water supply scenarios.

Basis for model development

The complexity of primary production systems and the need to fulfil multiple objectives call for a systems approach to better understand the chain of production processes (Kropff *et al.*, 2001). Such a systems approach is

---

[4] Xinyou Yin, Jan Vos & Egbert A. Lantinga

process-based ecophysiological modelling. Three types of models were available. A brief qualitative evaluation of these models indicated that the model of Yin et al. (2001) appeared to be the most suitable since many ecophysiological processes are modelled in a balanced and interactive way, using robust algorithms. Furthermore, basic soil processes (water balance and both carbon and nitrogen dynamics in soils), important for optimising resource use efficiency, are also explicitly included. So the starting point for model development was the structure of the model from Yin et al. (2001) and some elements from one of the other models (SUCROS; Goudriaan and Van Laar, 1994). In addition, elements absent in the existing models but specific to leguminous crops had to be incorporated, e.g. symbiotic nitrogen fixation (Jeuffroy and Ney, 1997). Another important new modelling part, especially with respect to the PROFETAS programme, is seed protein production.

Model development

To accurately predict crop yields, individual physiological growth-determining processes have to be adequately quantified. With respect to such processes, three major developments achieved in this project are: (1) a determinate growth function (Yin et al., 2003a), (2) generic relationships between leaf area index and canopy nitrogen (Yin et al., 2003b), and (3) a new equation for electron transport in leaf photosynthesis (Yin et al., 2004). Further, modelling individual physiological processes have been elaborated for (1) nitrogen fixation, (2) root senescence in analogy to leaf senescence, (3) the formation and remobilisation of stem and root carbon reserve pools, and (4) seed protein production from the amount of nitrogen partitioned to seeds. We also adopted new methods, reported in recent literature, for simple mechanistic modelling of canopy photosynthesis and crop respiration.

Integration of these individual model components resulted in a new, innovative generic crop growth model GECROS, Genotype-by-Environment interaction on CROp growth Simulator (Yin and Van Laar, 2005). The conceptual structure of the model is shown in Figure 3-4. The model has been coupled to a general soil model for assessing the dynamics of a crop-soil continuum. Required inputs include latitude of the site, daily minimum and maximum temperature, "global" solar radiation, vapour pressure, wind speed, rainfall, and air CO_2 concentration. The other environmental inputs are water availability and nitrogen supply from soil, which are predicted by the soil model. The model is applicable to any production level (either potential, or water-limited, or nitrogen-limited) free of pests and diseases.

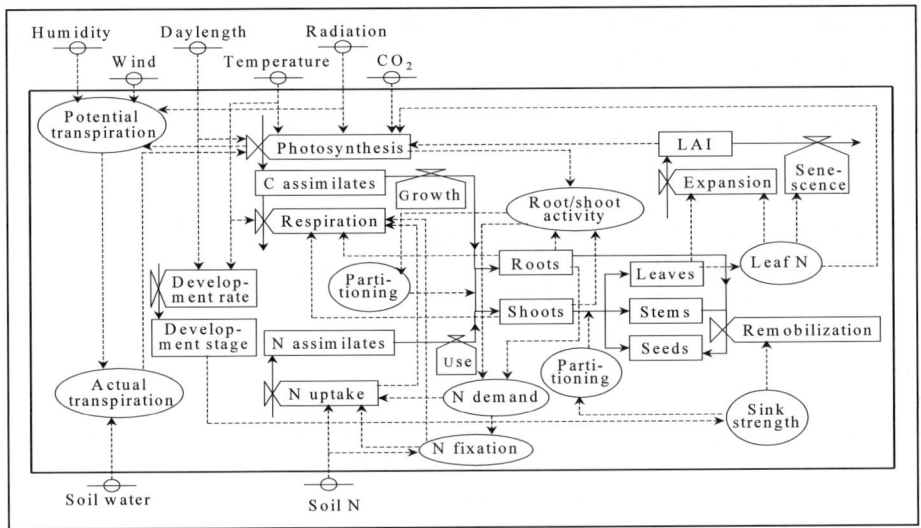

Figure 3-4. The relational diagram of the crop growth model GECROS (Abbreviations: C – carbon, N – nitrogen; LAI – leaf area index).

Model evaluation

For model evaluation eight pea cultivars (Solara, Classic, Santana, Athos, Phonix, Supra, Espace, Attika) were tested. These cultivars were chosen because they are currently grown as important pea genotypes in different parts of Europe. A glasshouse experiment was performed for estimating genotypic coefficients in the GECROS model, such as phenological parameters in these cultivars.

Performance of the GECROS model in predicting pea production under field conditions was assessed by using the same eight cultivars. They were sown on three dates. For the first and third sowing dates, 70 kg nitrogen ha^{-1} was applied (peas are commonly supplied with a small amount of fertiliser nitrogen in the early growth stages when nitrogen fixation capacity is being established). For the second sowing date two basal nitrogen levels (0 and 70 kg nitrogen ha^{-1}) were included. In this way four different growth environments were created. The experiment did not show a consistent, significant effect of applied nitrogen in these cultivars, probably because "native" soil nitrogen was sufficient to sustain early growth.

Tested with the data from these experiments the model appeared to perform adequately in explaining environmental effects on these genotypes (Figure 3-5). However, because of the unexpected similarities in genotypic characteristics among our tested cultivars and the low resolution of field data, we cannot assess at this stage whether the GECROS model can distinguish genotypic differences in final complex traits such as seed biomass and protein yields.

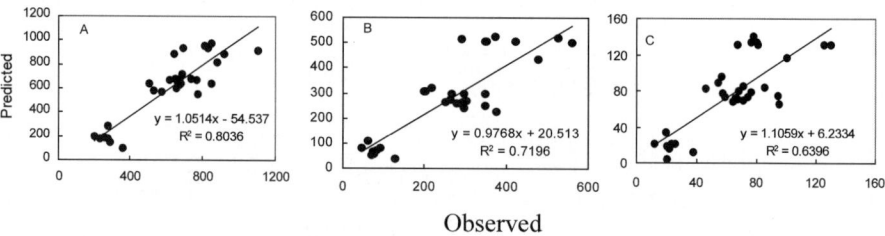

Figure 3-5. Simulated versus observed shoot biomass at maturity (A), seed biomass (B) and seed protein (C) (all in g/m²) in eight pea cultivars in three sowing-date growth environments (corresponding to three clusters of points).

Model simulation to assess pea production in Europe

The model was applied to a range of European conditions for pea crops, based on parameters for the cultivar "Solara". This cultivar was chosen because of its popularity across Europe and its use in other PROFETAS projects. Since nitrogen is usually not a limiting factor for the pea, we ran the model with three water supply scenarios (supply as crop demand, 200 mm and 100 mm initial soil available water), equivalent to production levels of ample water supply, loamy clay soil without irrigation, and sandy soil without irrigation, respectively. Simulations use climate data (1991-2000) from the Environment and Sustainability Institute of the European Commission for 66 pre-selected locations in Europe. Using the IDW (inverse distance weight) interpolation method with four nearest points, maps for average predicted seed yield and seed protein production of the ten years were made using a GIS software, under the three water supply scenarios (Figures 3-6 and 3-7).

Not surprisingly, predicted crop productivity depends strongly on water supply scenarios for all sites. Yearly variability in predicted crop productivity was greater under water-limited conditions. Areas with potentially high predicted productivity, such as Scotland, Denmark, North Germany, and part of France, are also regions in Europe where peas are currently widely grown. The Netherlands seem to be well suited for growing peas. The higher productivity in North Western Europe and Southern Scandinavia than in Southern Europe was basically due to longer crop growing periods due to cooler environments in the north.

However, caution should be taken, since the maps in Figures 3-6 and 3-7 were drawn without considering information about geographic landscape and soil quality. Furthermore, our simulation was conducted for only 66 sites, and in some areas (such as Scandinavia) mapping was merely the result of extrapolating a few point simulations. Higher-resolution maps could be made with more extensive simulations that incorporate local specific soil and landscape information.

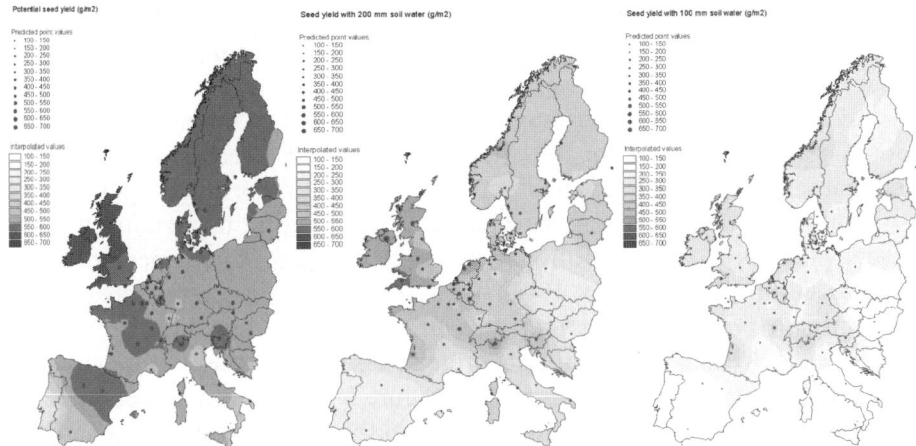

Figure 3-6. Map of pea seed yields under three water supply scenarios from interpolation of point simulation for 66 sites (points).

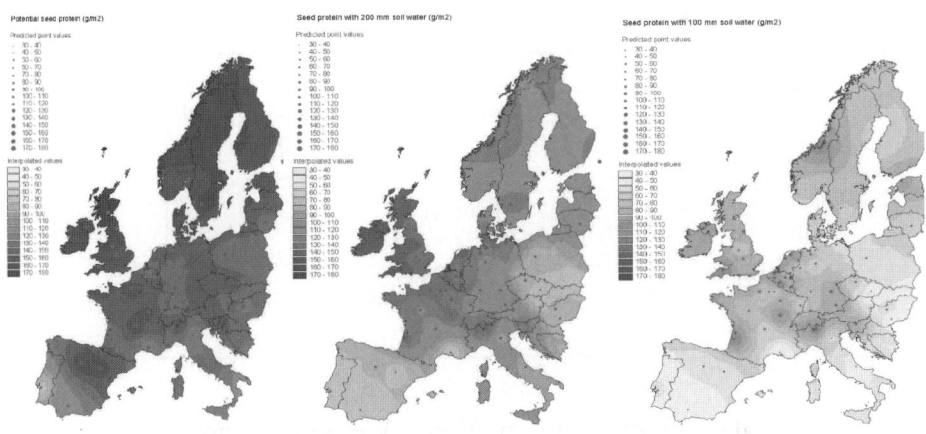

Figure 3-7. Map of pea seed protein under three water supply scenarios from interpolation of point simulation for 66 sites (points).

3.4.3 Concluding remarks

On the basis of our work on quantification of individual physiological processes, an innovative crop growth model, called GECROS, was developed. This model is not applicable to peas only, but it is generic, applicable to any arable grain/seed crop at any production level. GECROS requires only an input of a small number of parameters, which, in general, can be readily obtained. In addition to yield traits that most existing crop models predict, quality traits such as seed protein content are also predicted by GECROS. To avoid over-interpretation, it should be emphasised that the model does not account for the effects of pests and diseases on production and that it requires further validation by performing more field experiments.

From a PROFETAS point of view, it is noteworthy to mention that the model predicts – within the reported range of seed protein percentage in existing germplasm (Cousin, 1997) – that there is no chance to increase total seed protein production per unit area (the maximum being 1.8 ton/ha) by using pea cultivars having a high protein content, because these cultivars would have lower seed biomass yields (Figure 3-8). The lower seed yields of high protein cultivars were predicted largely because per unit of time these cultivars would need a higher amount of nitrogen to be withdrawn from vegetative parts in order to support the synthesis of the high amount of seed protein; the faster nitrogen withdrawal from vegetative parts causes faster leaf senescence and shortened crop photosynthetic duration.

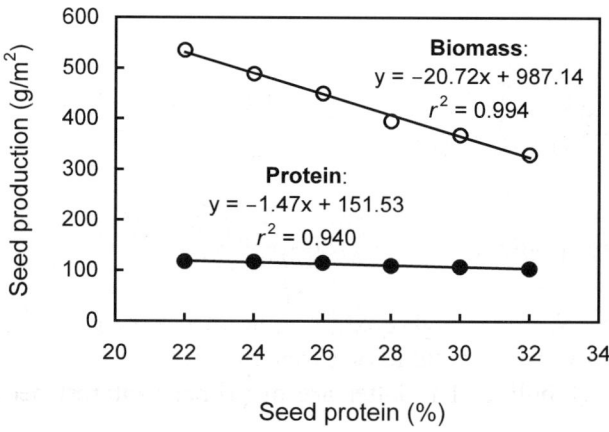

Figure 3-8. The relationship of seed protein and biomass yield versus seed protein percentage in pea crops as simulated by the GECROS model.

In practice, pea performance appears to be sensitive to excess water or drought during flowering and harvesting. The pea easily lodges in heavy rains, for instance, presenting a major problem to harvesting and a major risk to fungal attack. Improved straw stiffness has been and still is a major focus in pea breeding. The effect of drought and lodging severity in reducing canopy photosynthesis and seed set can be well assessed by GECROS. However, it has been beyond the reach of the current project to rigorously quantify the effect of excess water and lodging incidences per se, because of the lack of data. Soil-borne fungal diseases are a second practical problem of the pea. Root rot diseases, in particular "near wilt", caused by the fungus *Fusarium oxysporum* Schlecht. f. sp. *pisi* (J.C. Hall) W.C., needs to be mentioned. Since no cure exists, prevention is the only measure that can be taken. The best prevention is to grow a crop of peas on a field only once every six years.

As previously stated, GECROS is not only applicable to peas or to assessment of production levels. The GECROS model could also be a powerful tool in (1) predicting responses of global environmental change on crop production and cropping systems, (2) defining crop ideotypes adapted to a target environment, (3) optimising management strategies for specific crop genotype and environment, and (4) designing sustainable cropping systems. More specifically with respect to programmes such as PROFETAS, GECROS can also be used to analyse and identify the difference among various (leguminous) crops in resource (such as water) use efficiency in terms of biomass, protein and starch production and to assess quantitatively the consequences of an increased cultivation of (leguminous) crops on soil fertility and environmental load.

3.5 BREEDING: MODIFYING THE PROTEIN COMPOSITION OF PEAS[5]

3.5.1 Introduction and research goal

Peas are a relatively high-quality source of proteins. As mentioned in Section 3.1.3, the water soluble or extractable pea proteins are divided into albumins and globulins. The latter are of primary interest because of their

---

[5] Emmanouil N. Tzitzikas, Jean-Paul Vincken, Krit Raemakers & Richard G.F. Visser

large abundance and, hence, their dominant role in applications. The pea globulins can be subdivided into three major classes: legumin, vicilin and convicilin. However, pea globulin heterogeneity is much larger than this classification may suggest. This is due to a number of factors including: (i) the proteins of each class are encoded by small gene families leading to different isoforms, (ii) differential modification of the proteins after their translation from the encoding mRNA, both differential proteolytic processing (mainly vicilin) and differential glycosylation (mainly vicilin). Research on other legume seeds (e.g. soy) has indicated that the various isoforms and modified proteins have different functional properties, and therefore potentially different behaviour in industrial processes. This was recently also exemplified for the pea by the finding that the gelling behaviour of protein mixtures, analogous to those commercially produced, is affected by the cultivar from which the proteins are isolated (Section 3.3; O'Kane et al., 2005), and that convicilin has a negative effect on the gelling behaviour of pea proteins at near neutral pH (Section 3.3; O'Kane et al., 2004b).

In order to obtain pea seeds with a protein composition that is desired for producing NPFs, we have followed two approaches. First, the natural variation in protein composition was explored to select for pea lines enriched in one of the protein classes. Second, it was anticipated that natural variation would probably not provide for the desired specific protein composition, due to the fact that the pea proteins are encoded by multi-gene families. To facilitate the acquirement of a more uniform protein composition, we embarked on genetic modification. It is a problem that the pea is rather recalcitrant to transformation, and that the existing protocols seemed inefficient and laborious. Therefore, we started to develop a new transformation system for the pea.

3.5.2 Natural variation in pea protein composition

A collection of seeds from 54 pea lines, as well as from five pea wild relatives (taxa) (*P. sativum abyssinicum, P. arvense, P. elatius Marbre, P. elatius 1140175, P. elatius 1140176*), were subjected to protein compositional analysis by SDS-PAGE / densitometry (both under reducing and non-reducing conditions) after extraction of ground pea seeds with 0.1 M Tris at pH 8 (Tzitzikas et al., 2005b). The collection includes wild-type (normal) and *afila* (semi-leafless) leaf shape varieties. The seeds have different external characteristics such as size, shape and seed coat colour.

Globulins were the major category of seed proteins in all lines tested, except for one, their content ranging between 49% and 82%. No significant difference in globulin content was observed between the *P. sativum* species and the wild relatives. Vicilin is the most abundant protein of the pea and its content varied between 26% and 52% of the total extracted protein. In all tested material, both proteolytically processed and non-processed vicilins were found, the former being the predominant of the two, ranging from 18-41%; the non-processed vicilins constituted between 3% and 14% of the total extracted protein. Generally, convicilin was the least abundant globulin with a content ranging from 4% to 8%. The legumin content varied between 6% and 25% of the total extracted protein. The wild relatives, in general, contained more legumin (20%, on average) than the *P. sativum* lines (14%, on average). Next to globulins also proteins related to globulins appeared to be present in the extract, in amounts ranging from 3% to 17% of the total extracted proteins. The extremes in protein composition are summarised in Figure 3-9. It is clear that pea lines can vary considerably in their globulin composition, although none of the lines lacked any particular globulin. We have not observed apparent correlations between a high content of one globulin and a low content of another one. The ratio of vicilin/legumin varied from 1.2 to 8, and is significantly higher in *P. sativum* than in the wild relatives, with averages of 2.8 and 1.7, respectively. No significant correlations between this ratio and external characteristics (such as leaf shape, seed shape, and seed colour) were found. The vicilin/convicilin ratio varied from 3.4 to 11.5, and appeared significantly lower in the *afila* lines than in the wild type, with averages of 5.6 and 6.8, respectively. No differences between the wild relatives and *P. sativum* lines were observed.

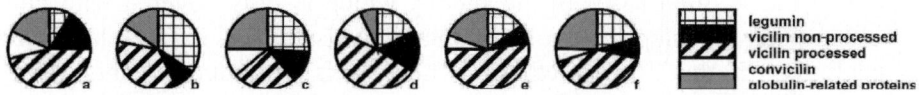

Figure 3-9. Examples of extremes in globulin composition in various pea lines. For legumin: a = P. sativum NGB 102149 Iran, and b = P. arvense CGN 10193. For vicilin: c = P. sativum CEB 1162, and d = P. sativum 93125-07 Canada. For convicilin: d, and e = P. sativum Cisca. For globulin-related proteins: d, and f = P. sativum NGB 102920 Iran.

Of particular interest was the observation that the extent of proteolytic processing of vicilin can differ considerably between lines; the ratio of processed to non-processed vicilins (pV/npV) varied from 1.6 to 8, showing that the processed fraction is always higher than the non-processed one. Significant differences were found between lines with different leaf shape and seed coat colour with respect to the pV/npV ratio. The average ratios of *afila* and wild-type lines were 2.9 and 3.9, respectively, whereas that of brown-grey spotted seed-coat lines (*P. elatius* accessions) with 6.3 was significantly higher than the maximum of 3.5 of the other lines with different seed-coat colour.

3.5.3 Development of a regeneration and transformation protocol

The pea, like most legume species, is recalcitrant to genetic modification. The systems available (Schroeder *et al.*, 1993; Grant *et al.*, 1995; Bean *et al.*, 1997) are all based on using meristems of germinated seeds as starting material. After infection with *Agrobacterium tumefaciens* the germinated seeds are cultured on a medium where they form multiple shoots, some of which are genetically modified. However, the existing genetic modification protocols have drawbacks such as a low efficiency (only about 1% of the initial seeds is (partially) transformed), high frequency of "escapes", and the "chimeric" nature of the transgenic plants. Therefore, these systems are not used routinely for the production of large numbers of transgenic plants. The disadvantages might be related to the non-adventitious nature of the regeneration system of transgenic plants, and it was hypothesised that they might be reduced significantly if an adventitious regeneration system would have been used. Such a system was developed, which is essentially a 5-step procedure. In step 1, stem tissue (consisting of one node) from *in vitro* plantlets is cultured on a medium where multiple shoots are produced, either normal looking or swollen at their bases. The multiple shoots are subcultured in the same medium, resulting in the formation of a green "hyperhydric" callus in the swollen bases of the shoots, which is fully covered with bud-like structures. In step 2, this bud-containing tissue (BCT) is isolated and subcultured in the same medium, where it is reproduced. This newly reproduced BCT material has a round shape, and can be subdivided and subcultured in the same medium in a cyclic fashion. In step 3, BCT is subcultured on a medium supplemented with a combination of specific hormones to form shoots. In step 4, the addition of a hormone for root formation to the medium resulted in normal looking plants with roots. In step 5, *in vitro* plants are transferred to the greenhouse for acclimatisation and further development. This procedure was tested with four varieties (Espace, Classic, Solara, and Puget), which were all able to produce the BCT, suggesting that its production is genotype-independent.

The regeneration system outlined in Figure 3-10 has been used to produce transgenic plants. In the first 2 years of experimentation a construct containing only an easily detectable gene (the luciferase gene) was used. In these initial experiments, BCT (stage c in Figure 3-10) was infected with *Agrobacterium* strain AGL1, and then cultured on a medium for BCT multiplication and enrichment in transgenic tissue. These experiments were not very successful; either the transgenic tissue was lost after a few subcultures, or the tissue died. In fact, 2 years of experimentation resulted in only three genetically modified lines from which one plant was regenerated. This plant was transferred to the greenhouse. Although this plant did not grow very well, it produced 2 seeds after selfing, one of which was luciferase positive. The plant derived from this seed was much more vigorous than the original plants, and produced a total of 30 seeds, of which about 50% were luciferase positive.

In the final year of experimentation another construct was used which contained the luciferase gene together with an interrupted inverted repeat (IIR) of the legumin A (*LegA*) gene (an IIR construct is a more efficient tool for down regulation of the amount of a target protein than the conventional antisense constructs). Different strategies were tested to produce genetically modified plants. Only one of these strategies led to the production of genetically modified tissue: infection of shoot-containing BCT (Figure 3-10d) with *Agrobacterium*, and subculturing on a medium for BCT multiplication followed by culturing for plant generation (performing step 6 and subsequently steps 2 to 5 as indicated in Figure 3-10). This strategy resulted in the production of more than 50 luciferase positive BCT lines. Twenty of these were cultured for plant regeneration. At the moment, plants of 10 lines are grown in the greenhouse. Also with this new construct, the plants do not grow very well in the greenhouse, just like those with the construct containing only the luciferase gene. One line has already produced luciferase positive seeds, indicating that the construct is transferred to the next generation. Currently, from these seeds plants are grown for the production of the second generation of transgenic seeds. In the near future, also the other genetically modified lines will produce seeds.

Figure 3-10. A 5-step procedure for the initiation of bud-containing tissue (BCT) in pea (Pisum sativum L.), and subsequent regeneration of plants. Different steps are indicated by numbers and explained in the text. a: pea plant obtained from seed with the one-node explant encircled; b: production of swollen multiple shoots with at their base meristematic BCT (indicated by the circle); c: fragments cut from BCT can be subcultured, resulting in their reproduction in a cyclic manner; d: the formation of shoots from BCT; e1: shoots are harvested and subcultured; e2: shoots are rooted; f: BCT-derived plants are grown in the greenhouse.

3.5.4 Concluding remarks

This research has shown that natural variation of pea protein composition is large, but that lines lacking one of the protein classes do not seem to exist. An important finding is also that there is considerable natural variation for proteolytic processing of vicilin. Although the relevance of this for technological applications needs to be further established, we think that this post-translational processing may influence the aggregation behaviour of vicilin. Until now, plant breeders have largely overlooked this trait. For more extreme variation in protein composition genetic modification may provide an alternative, particularly now that an improved regeneration and transformation protocol has been developed for the pea (Tzitzikas *et al.*, 2004; Tzitzikas *et al.*, 2005a; Shan *et al.*, 2005b). This new protocol is applicable to different pea genotypes. All transgenic plants transfer the transgenes to the next seed generation. We have shown that other legumes such as soy respond in a comparable way to the new regeneration procedure developed for the pea (Shan *et al.*, 2005a). Also these crops are recalcitrant to genetic modification, and our regeneration and transformation system might offer opportunities to improve the transformation efficiency of these species.

We have now generated a number of transgenic pea lines, in which we aim to alter the globulin composition of the seeds. More in particular, we have embarked on reducing the legumin content by introducing an interrupted inverted repeat of the *LegA* gene. Currently, seeds are produced by approximately ten lines. In the short term, it remains to be established (i) how far the legumin content can be reduced, (ii) whether only one legumin isoform (the one encoded by the gene used to make the construct) is affected, or several isoforms belonging to the legumin class, and (iii) whether there is a yield penalty for total protein content, in other words whether a reduction in legumin leads to an increase in the amount of other proteins such as vicilin and convicilin. Further, the germination power of the seeds needs to be determined, since it is unknown whether modification of the seed reserves affects these characteristics. In the longer term, we also intend to generate pea lines that are down-regulated in the other globulins, and in which post-translational processing of vicilin is reduced. In collaboration with food science research groups, the various transgenic seeds will be assessed for differences in functionality.

3.6 METHODOLOGY FOR CHAIN DESIGN[6]

3.6.1 Introduction and research goal

A supply chain in general is a sequence of events carried out by various actors leading to a product or a service. It can be viewed as an integrated process where raw materials are acquired, converted into products and then delivered to the consumer (Beamon, 1998). The chain is characterized by a forward flow of goods and a backward flow of information and money. Supply chain design and optimisation can be seen as a process or a procedure to make this sequence of events as effective and as efficient as possible. Effective in this context means that the goals are indeed achieved and efficient that the goals are reached at as low a cost as possible.

The main fact that differentiates food supply chains from other chains is that there is a continuous change in quality from the time the raw materials leave the grower to the time the product reaches the consumer (Tijskens *et al.*, 2001). Food supply chains are made up of organisations that are involved in the production and distribution of plant and animal-based products. Such supply chains can be divided into two main types (Van der Vorst, 2000):
- Chains for fresh agricultural products: the intrinsic characteristics of the product remain virtually unchanged, i.e. an apple remains an apple and,
- Chains for processed food products: agricultural products are used as raw materials to make processed products with a higher added value, e.g. apples are converted to applesauce.

The NPF supply chain clearly belongs to the second type of chain. The NPF chain as defined in this book consists of six links: primary production, ingredient preparation, product processing, distribution, retail and the consumer (Figure 3-11). As indicated in this figure, each link may be made up of more than one actor. So there are many actors involved in the NPF supply chain. Each actor in this chain may optimise his own performance. However, such a situation may not be optimal for the complete chain. The concept behind chain design is that the chain is looked at in its entirety, i.e. it aims at the performance of the chain as a whole. During chain design, parameters that can be manipulated are the flows of the products, the processes they undergo, the choice of where to handle the raw materials, product, etc. The performance criteria as such can be many, for example the production cost, texture, taste, land used to make the final product and

[6] Radhika K. Apaiah

energy input. This complicates the design story. In view of this complexity, this study aims at developing a methodology for chain design, thereby aiding in identifying feasible options to produce a pea-based novel protein food.

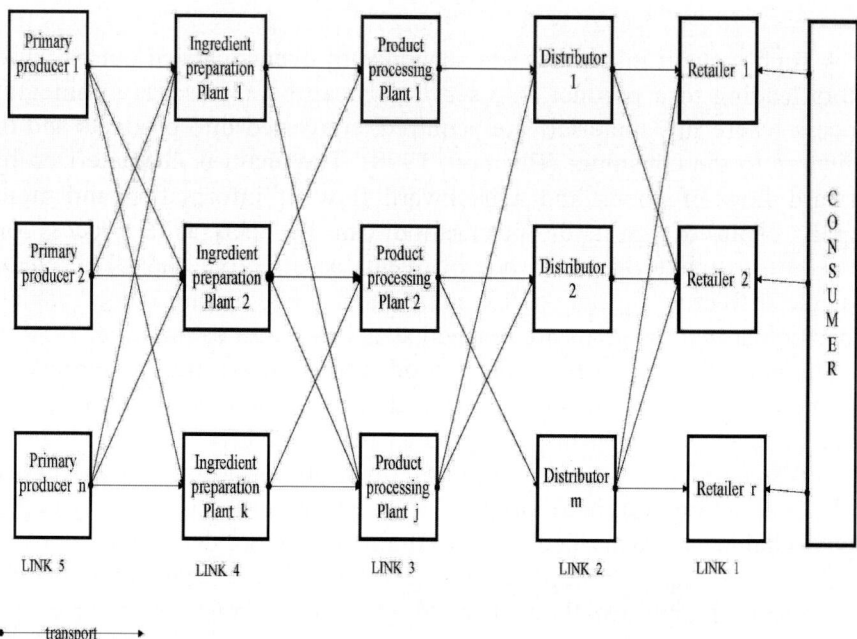

Figure 3-11. Food supply chain (Apaiah et al., 2005a).

3.6.2 Methodology

First, a qualitative model of a supply chain is developed. For a given product, attributes, goals and performance indicators for the goals are selected (see Table 3-1 for examples). The production chains are chosen and the relationships between the performance indicators and the control variables are determined in order to develop a quantitative model. Figure 3-12, the qualitative model, illustrates the steps that should be taken by the chain designer according to this methodology.

TECHNOLOGICAL FEASIBILITY

Table 3-1. Examples of attributes, goals, definitions and performance indicators of a supply chain.

Attributes	Goals	Definition	Performance indicators
Good taste	Good quality	Texture	Water holding capacity
		Flavour	Sugar, flavour volatiles
		Colour	Carotenoids, chlorophyll, Maillard reaction products
		Nutritive value	Anti nutritional factors, essential amino acids, cholesterol
Cheap	Low cost	Low selling price	Euro/kg product
		Low cost price	Euro/kg product
Does not harm the environment	Low environmental load	Low energy use	Joules/kg product
		Less waste	Kg waste/kg product
		Exergy	Exergetic efficiency

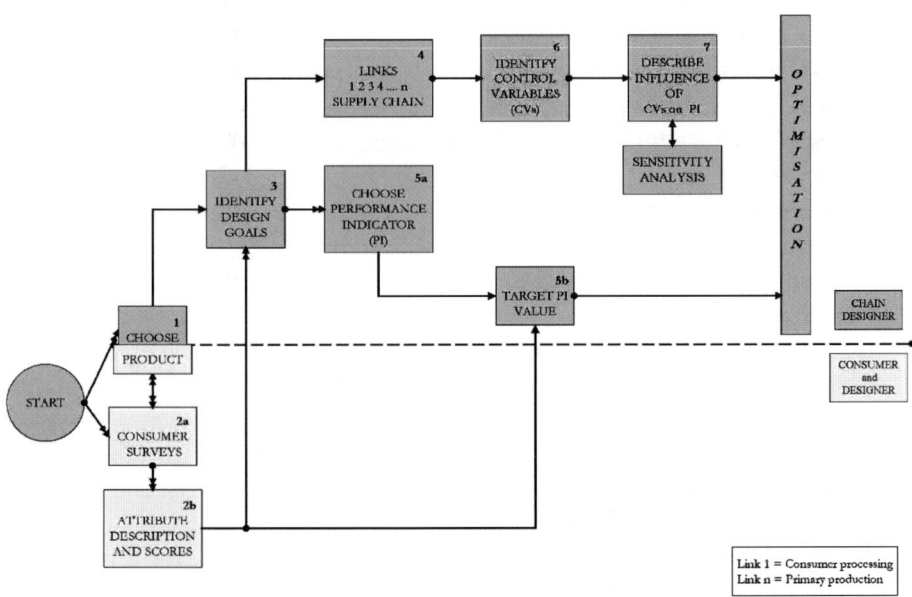

Figure 3-12. Methodology to derive a qualitative model.

Applying this methodology, three separate studies were conducted, each aimed at optimising the supply chain towards a different goal: quality, cost and environmental load. Since this project is focused on methodology development, rather simple indicators were chosen for these parameters. Quality is defined as good texture, which is related to water holding capacity of the protein in an NPF; this is admittedly a gross simplification of quality, but it serves its purpose for methodology development. The same applies to cost. The environmental load is evaluated by analysing the exergy (see below for a definition) requirements for the NPF chain (Apaiah *et al.*, 2005b). Two optimisation techniques were used: non-linear programming for process optimisation (for quality and for environmental load of the product) and linear programming for flow optimisation (best path for lowest cost).

After the challenges and issues relating to these three goals have been investigated separately, they will be looked at together, i.e. will be integrated. This can be done by techniques like multicriteria analysis, where many alternatives are generated and ranking techniques are used.

3.6.3   Results

1) Optimisation towards quality

A study on the design of an NPF supply chain with respect to the quality aspect led to the following insights (Apaiah *et al.*, 2005a):
- A systematic methodology has been set up to study supply chain design.
- The choice of the performance indicator is very significant as this traces changes in the goal as the product moves through the chain. Extensive literature research and contact with experts showed that it is extremely hard, but possible, to detect all relevant relations between the performance indicators and the control variables in the chain.
- All the links in the chain from primary production up to and including consumer processing are relevant, but the influence of the links on the final product change with the goal.
- Chains have to be designed and optimised for a specific product as the chain pathway changes with the product.
- Chain design also changes with the goal.

2) Optimisation towards product costs

A second study on the design of a supply chain network for a pea-based NPF was undertaken to investigate chain design when lowering product cost is the main goal (Apaiah and Hendrix, 2005). The study focused on finding the lowest cost at which NPFs can be manufactured for a specific market demand. The decisions are related to the location of (a) primary production, (b) ingredient processing and (c) final NPF production areas, as well as modes of transportation. Finally, the sum of production and transportation costs is minimised. The model that was developed was used to generate scenarios based on practical constraints such as capacity limits. The cost estimations were limited to those for the main ingredient, the pea, and the concentrate made from it. The procurement costs for the other ingredients like oil, functional ingredients, flavours were not considered here. It is known, however, from preliminary calculations, that the inclusion of these costs would still limit the production cost per ton of product to below € 1,000. Further value is added with the inclusion of the costs of the second part of the supply chain, namely packaging, distribution and retail. NPFs are targeted to replace pork in the consumer's diet. The retail cost of pork is about € 6/kg and the cost to produce pork is about 40% of this value. Evidently, NPFs production costs of, say, € 1/kg would compare favourably with pork at about € 2.40/kg.

3) Optimisation towards environmental load

This study explored the potential of using exergy analysis to study and compare the environmental impact of food supply chains (Apaiah *et al.*, 2005b). Exergy is the available energy or available work or quality of energy of a source, i.e. it is a measure of the work that can be done by a source. An exergy balance when applied to a process or a whole plant, shows how much of the available work potential supplied as the input to the system under consideration has been consumed (irretrievably lost) by the process (Kotas, 1995). It enables the determination of the location, types and magnitudes of wastes (streams that still contain exergy) and losses (exergy is irreversibly lost). Exergy analysis of the chain identifies the links where exergy loss destruction takes place and shows where improvements are possible to minimize this loss.

The supply chains of three products were analysed: minced pork meat, an NPF made from dry peas and pea soup. Exergy content and requirements of the various streams, products and processes were calculated for the three chains. As exergy is expressed in one unit, the Joule, the inputs and outputs of each chain are easily comparable. The contributions of the links to the total exergy loss are different in each chain. As is illustrated in Figure 3-13, in the NPF chain the greatest input is required in the processing link, i.e. ingredient preparation and product processing. This processing link is exergy intensive, because the peas have to be sorted, dried to reduce moisture, dehulled, milled, fractionated to a protein preparation (ingredient preparation) and converted to NPFs, for example, by extrusion (see 3.1.2), a process that is not required in the pork chain. In this chain primary processing and transportation require the highest inputs. The former is due to the inefficient conversion of plant protein from the feed to animal protein in the pig and the latter to the required cooled transport.

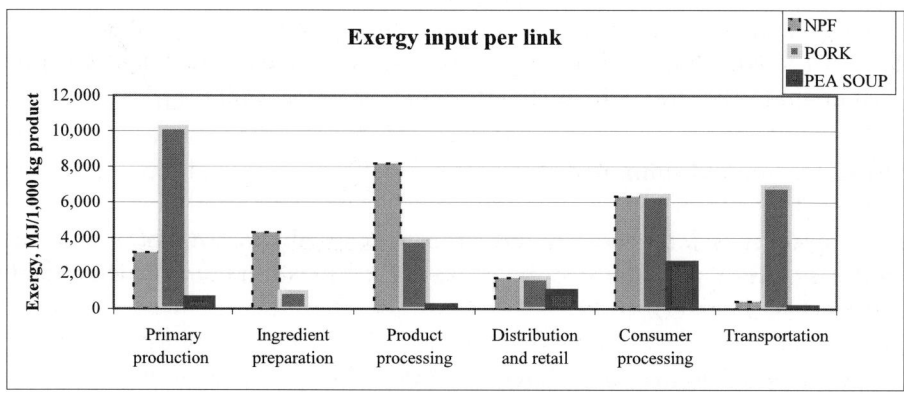

Figure 3-13. Comparison of exergy input per link between the three chains.

Figure 3-14 compares the exergy required for the three chains with respect to exergy input per stream. The high exergy input in the pork chain due to drinking water requirements is striking. In the NPF chain, the input of ingredients requires more exergy than in the other two chains due to the exergy intensive preparation of pea proteins (see above).

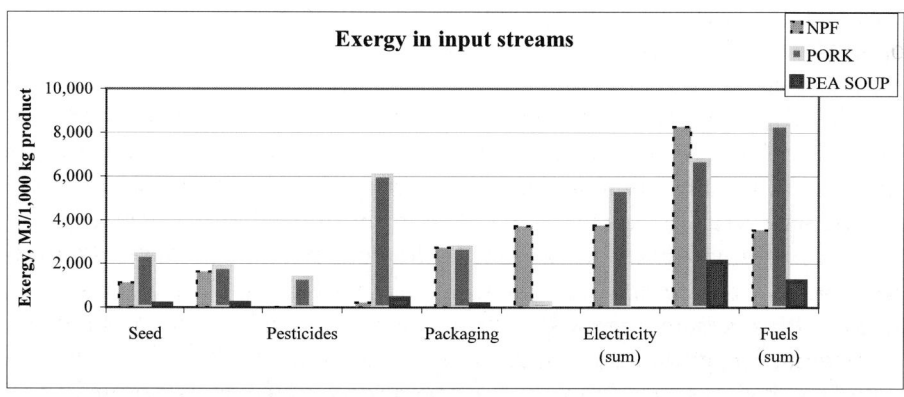

Figure 3-14. Comparison of exergy input per stream between the three chains.

The exergy analysis showed that the NPFs chain (excluding the non-protein part of the crop) is only slightly more efficient (1.2 times) than the pork meat chain. Both chains require much more exergy than the pea soup chain. However, only part of the input exergy goes into the desired product, the remainder goes into waste streams or is irreversibly lost to the environment. The largest output in terms of exergy in the pea soup chain is the desired product, whereas in the NPF chain it is the raw material waste, especially the starch enriched fraction (see Section 3.7), which is much larger than the protein enriched fraction that is used for NPF production. The desired product, the NPF, is second by far. The largest output in the pork meat chain is manure with meat only 15% of the total output.

3.6.4 Conclusions

A systematic methodology has been developed to design supply chains for NPFs. Such a systematic approach helps to structure the demand for relevant data and can generate interesting alternatives based on certain constraints. Studies were conducted to design chains aimed at optimising quality, minimising costs and environmental load, respectively. The design of the chain changes with the specific goal and with the particular product. Initial cost calculations show that NPFs are viable, their prices can be less than or at least comparable to those of pork meat. Furthermore, it was found that in terms of exergy, the NPF chain is only slightly more efficient than the pork meat chain. However, it should be emphasized that if the waste stream (mainly the non-protein fraction; see the next section) is utilised (and it is evident that such is a prerequisite for transition), the efficiency of the NPF chain in terms of exergy will strongly increase because most of the exergy goes into this stream. Exergy analysis was found to be very useful to recognise problem areas within the chain and to identify the potential for optimisation of the chain with respect to environmental impact (see Chapter 2 also).

The first three studies looked at the goals for chain design independently of each other. However, in the real world choices or goals are not that simple. More often than not, decisions have to be made when many possibly conflicting goals or criteria have to be considered simultaneously. The problem is further complicated because of the presence of many alternative pathways for NPF production that need to be evaluated. In the context of this study, it means finding a method to design supply chains with multiple goals or criteria, i.e. cheap, good quality and low environmental load. Multiple criteria decision analysis is a decision making tool that is useful in these circumstances and will, therefore, be used to develop the methodology to design such chains.

3.7 OPTIONS FOR NON-PROTEIN FRACTIONS[7]

3.7.1 Introduction and research goal

To enable a transition from meat to plant protein, it is important to consider what options exist for the non-protein fraction. Any crop that is used to produce NPFs will, at one or more points in the production chain, also yield a fraction that cannot be used for the NPF production. These residues may arise in various forms and may have different compositions depending on the crop used, the part of the production process in which they arise, and the actual production techniques used. This means that both the economic value and the environmental impact of the non-protein fraction can vary tremendously.

The goal of the research presented here is to identify what non-protein fractions may be produced, the possibilities that exist for making use of non-protein fractions and what environmental pressures are associated with these uses.

3.7.2 The non-protein fraction

To study the non-protein fraction, information is needed on the constituents of any crop that is used for its protein. Of the main commercial food and feed crops the main constituents other than protein are oil and carbohydrates such as starch. For practical purposes, we will primarily consider oil and starch. It should be kept in mind, however, that in reality crops are more complicated than that. For some crops it is common to harvest the residues, such as straw, which are rich in cellulose, and use these residues for feed (ruminants) or for non-food applications such as paper manufacturing. Cellulose could also play a role if one were to look at more unusual sources of plant protein, such as grass. Besides cellulose, many common food and feed crops contain other non-starch polysaccharides (NSP), which currently have few known large-scale uses other than feed. Lastly, in many kinds of crops interesting chemicals may be present in relatively low quantities in the non-protein fraction and in some cases these are commercially extracted and marketed. Chemicals in this last category, however, are not considered here.

---

[7] Frank Willemsen

For each constituent of the non-protein fraction there are different options. For current commercial food and feed crops, the options are likely to be food, feed, industrial raw material (termed stock in this section) and energy, whereas for the crops that leave a cellulose or NSP-rich fraction food application is not an option, but the other three uses are. For all options estimations can be made with regard to their economic value, in order to be able to judge how realistic any given scenario is. As a first approximation one can assume that the value of a certain use follows a decreasing scale in the order: food>feed>stock>energy. Apart from being able to give an indication of how realistic a certain option is, economic value is also important when it comes to allocating environmental impacts to the different fractions of a given crop. This is the case because it is quite common in environmental assessments (e.g. Life Cycle Assessment) to allocate environmental impacts to two or more products produced in one production process, using the economic value of each product as a measure for how much of the environmental impacts should be allocated to that particular product.

The uses of the non-protein fractions are influenced greatly by the way the protein is separated. Although it is technically possible to extract very pure protein fractions from practically any raw material, this is often not done in commercial practice, because it is economically not feasible or environmentally undesirable, which usually has economic consequences as well. From both an economic and an environmental point of view, it is usually better to look for options with fewer processing steps, and to look for those steps that involve less of auxiliary substances such as water, even if this has a negative effect on the overall yield of the extraction process. This does depend, however, on the end use: a wet separation step may pose problems for use as fuel, but it may not be a problem for use as stock.

For pea protein, for example, there are two main ways of extracting protein (see also 3.1.3). One is dry milling followed by air classification, which does not have the highest separation resolution but is relatively cheap and clean, whereas wet processing can give very high purity starch and protein fractions, but carries higher costs for drying and produces a considerable waste water stream. Although in the current commercial setting the higher purity of the starch fraction, in particular, has led companies to adopt this process, we will assume a separation by dry milling and subsequent air classification, because of the environmental implications. The end products of this separation are two pea preparations (protein-enriched and starch-enriched) that both contain mixtures of protein and starch.

3.7.3 Assessing different non-protein fraction on the basis of their constituents

To assess different crops on the uses that are available for the non-protein fractions a tool was developed. On a scorecard the different constituents of the non-protein fraction are placed below one another and their possible uses are divided into the four categories mentioned above: food, feed, stock and fuel. An example for such a scorecard is given in Table 3-2 for an imaginary crop X, divided into constituents X_1, X_2, etc. For every constituent the four categories are then considered, and if considered suitable in a category, what product they may replace. In this assessment of the non-protein fraction the level of detail that is applied to this fraction is limited to its bulk constituents. Only an oil fraction, a starch fraction, and a "rest" fraction will be distinguished, with the "rest" fraction consisting of molecules such as non-starch polysaccharides (NSP) and non-extractable quantities of starches and oils.

In every use/replace combination various aspects can be addressed. Technical aspects must be considered, possibly leading to the verdict "unsuitable", if there are very high technical barriers. Likewise, economic aspects need to be taken into account, in which case unrealistically high costs could also rule out options.

Table 3-2. Non-protein scorecard for crop X. (NSP = non-starch polysaccharides; *Syngas and biocrude are both experimental "biobased" streams designed to resemble current outputs of the petrochemical industry. Syngas is a mixture of carbon monoxide, hydrogen and short chain hydrocarbons, which can be used for example as monomers, whereas biocrude is a mixture of more long chain hydrocarbons. Co-firing is a process in which organic materials are fed into furnaces of power plants along with the usual fuel such as coal.)

Type of fraction	Option	Food	Feed	Stock	Energy
Starch (X1 %)	main use 1	syrup	pig feed	syngas*	biocrude*
	replaces	corn syrup	maize	mineral oil	mineral oil
Oil (X2 %)	main use 1	cooking oil	chicken feed	cleaning agent	biodiesel
	replaces	sunflower oil	maize oil	palm oil	mineral oil
	main use 2			plasticiser	
	replaces			mineral oil	
Rest (NSP, other) (X3 %)	main use 1	unsuitable	cattle feed	syngas*	co-firing
	replaces			mineral oil	coal

On the environmental side, the same scheme would serve to find the best combination of environmental benefits. As the basis of the environmental assessment is the non-protein fraction, which may have several possible uses, each prospective use can be attributed an estimated environmental impact based on the total impact of the process (whole crop) and an estimated economic value. This can then be compared to the environmental impact of the substance it replaces.

The reality of the economic and environmental constraints on the separation process mentioned above means that in most cases a separation of the protein crop into three or four main products has little to offer in terms of usability of those fractions. This means that for most crops the most likely option would be a two-way separation, resulting in a single non-protein fraction, reducing the "scorecard" to a single row as illustrated using a starchy crop in Table 3-3.

Table 3-3. Non-protein scorecard for a starchy crop X.

Type of fraction	Option	Food	Feed	Stock	Energy
Starch	main use 1	syrup	pig feed	syngas*	biocrude*
	replaces	corn syrup	maize	mineral oil	mineral oil

If we then compare the fractions of different crops there are marked differences between starch-rich fractions and oil-rich ones as is illustrated by using peas, soy and rapeseed (Table 3-4).

Table 3-4. Non-protein scorecard for three crops.

Type of fraction	Option	Food	Feed	Stock	Energy
Pea starch	main use 1	limited	pig feed	syngas*	biocrude*
	replaces		maize	mineral oil	mineral oil
Soy oil	main use 1	cooking oil	chicken feed	cleaning agent	biodiesel
	replaces	sunflower oil	maize oil	palm oil	mineral oil
Rapeseed oil	main use 1	cooking oil	chicken feed	cleaning agent	biodiesel
	replaces	sunflower oil	maize oil	palm oil	mineral oil

In the case of pea starch, assuming air classification, the options for use as food are limited, mostly due to protein content (about 6%) and the taste. For wide-scale use of such a fraction in various food products it should have as little taste as possible, whereas pea starch tends to have a "beany" taste. Use as pig feeds should not pose any problem, although other starch sources have higher yields per hectare, usually. Use for industrial purpose could be impaired by the protein as several industrial uses would require a higher purity of starch. Although this higher purity could be achieved by wet processing, there are some disadvantages. This has mostly to do with the high cost of drying and the cost (and indeed environmental impact) of dealing with large amounts of effluent (Slinkard *et al.*, 1990). Use of the pea starch fraction for generating energy is a very low value option.

If one starts out with oily protein crops, the easiest separation is crushing; the process involves flaking, followed by extraction with a non-polar, low-boiling solvent. The oil fraction is already relatively pure. The oil of both soy and rapeseed (provided it is a low-erucic acid variety) has many uses. Currently these oils are used for all four of the categories food, feed, stock and energy.

In interpreting these figures there are a few important points to be considered when it comes to comparing oil and starch. Firstly, the amounts of oil or starch produced per kg of protein. This will be illustrated using the three crops mentioned above. For peas the amount of starch produced per kg of protein can be as high as 3.5 kg. In the cases of soy and rapeseed per kg of protein 0.57 and 1.3 kg of oil are produced, respectively. Because the yields of the three crops are roughly the same at around 3 t/ha, this means that when peas are used more land is needed to grow the same amount of protein. However, when wet-processing is used this picture changes. Then no protein is lost in the starch fraction and the land use per kg of protein is comparable to that for rapeseed, but still higher than for soy. The – rather spectacular – effect of this is discussed in Section 6.3. Secondly, the energy density of oil is around twice that of starchy fractions, so when use as energy source comes into play, 1 kg of oil has to be compared with 2 kg of starch. This higher energy density can also be important in the feed and stock route.

3.7.4 Concluding remarks

Without useful application for the non-protein fraction the replacement of meat protein by plant protein is simply not feasible. A reduction in meat production, however, would most likely result in reduction of feed crops grown in the world. The reduction in feed crops is much larger than the increase in protein crops that would be needed for NPF production. Due to the inherent inefficiency of feeding protein crops to animals this is always the case, no matter what kind of animal one considers. The kind of animal only determines how much land one saves by using the protein for NPF production. Therefore the non-protein fraction has two different faces. In the case of a reduction of meat protein produced and the concurrent reduction in feed crops that are – in part – oil crops, a drop in the production of oil will occur. If oil crops are used to provide the protein for NPF production, the non-protein fraction is a relatively high-value side product and the production would partly compensate for the drop in oil production that is caused by the decreased demand for feed. If peas were used to provide the protein for NPF production, the non-protein fraction obtained by milling and air classification is a relatively low-value by-product for which there would currently not seem to be a profitable market. Furthermore, this starchy fraction can not compensate for the reduction in the supply of vegetable oil described above. For the moment, therefore, oil crops seem preferable over peas as a source of protein for NPF production. This does not mean that a large-scale use of oil crops for NPF production is without problems. Many oil crops contain large amounts of non-starch polysaccharides. Under the current practice where the product that is left over after crushing is sold as a protein-rich feed this is not a major problem, because cattle can digest NSP, and pigs can partly digest them and excrete the rest without too many consequences. In NPFs for human consumption, however, NSP could have a negative effect on the product quality and they should therefore be removed, thus constituting a large waste stream of relatively low value.

If due to technological or other factors the possibilities for utilising a large starchy fraction were to increase, it is still questionable whether peas are the optimal crop to produce that starch fraction. Due to their higher yield and starch content, crops such as maize and wheat would become much more profitable if there were an increased demand for starch, and that might well offset their lower protein contents.

3.8 CONCLUSIONS[8]

The overarching goal of the technological projects was to answer one of the three main questions of the PROFETAS programme: Is the production of NPFs technologically feasible? The answer is affirmative and, in fact, a kind of NPFs, i.e. meat substitutes, are already available. However, their market share is quite low, most likely because they do not meet consumer preferences with respect to sensory quality. In order to reach a larger market share, the important question is: Is it technologically feasible to produce NPFs that better align with consumer wishes and preferences, in other words to produce NPFs that appeal to consumers so well that a protein transition may occur? With this in mind and taking into account the focus on pea protein, the main questions to be tackled are:

---

[8] Johan M. Vereijken & Martinus A.J.S. van Boekel

- *Breeding.* Is it possible to adjust the protein composition of peas towards the desired composition for the production of NPFs?
- *Cultivation.* What are the optimal cultivation systems with respect to protein yield and resource use efficiency for peas having the desired characteristics for NPF production?
- *Processing.* Is it possible to produce NPFs that better meet consumer demands with respect to flavour and texture than present meat substitutes?

As is summarised below, the technological PROFETAS projects have provided information concerning these questions.

Breeding project (Section 3.5)

There is a large natural variation in protein composition of peas, making it possible to breed via "classical" means for peas having a wide range in protein composition. However, no cultivars have been found that lack one of the major protein classes. This puts limits on the "classical" breeding approach to vary the composition. Complete elimination of, or a very large reduction in, the amount of one of these classes might be possible by the alternative tool that has been developed in this project, i.e. an efficient and effective protocol to genetically modify the pea by recombinant DNA technology. Besides varying protein composition, this protocol also allows removal or reduction in amount of enzymes (and other substances) that may interfere with quality demands for NPFs (e.g. the enzyme lipoxygenase that may be involved in the formation of "off-flavours" by oxidation of unsaturated fatty acids).

So the answer to the question whether it is possible to adjust the protein composition towards a composition that may be desired for specific NPFs is in principle affirmative, for tools have been provided. However, it is still unknown to what extent the protein composition and the amount of enzymes (or other substances) can be manipulated without affecting the viability and the yield of the peas in the field.

Cultivation project (Section 3.4)

A crop growth model has been developed by which the yield of peas and pea protein can be predicted in response to genotypic characteristics and environmental variables. This implies that for a given pea cultivar the model can be used to predict its yield in a particular environment, to optimise the cropping system with respect to resource use efficiency or to select the appropriate environment. Working backwards, the model can be used to define crop characteristics for a particular environment.

So the answer to the question about the optimal primary production system (in terms of seed and protein yield and resource use efficiency) for peas having the desired characteristics for NPF production is affirmative, since a model has been developed to define the optimal cultivation systems. However, the model still has to be validated.

Processing projects (Sections 3.2 and 3.3)

Knowledge has been obtained about the amounts of model flavour compounds that bind to pea proteins under various processing conditions, a valuable tool for flavourists to direct the taste of the product. Furthermore, it was shown that classes of pea proteins differ in their ability to bind flavour molecules, implying that differences in protein composition can be exploited to induce differences in taste of NPFs. In addition, "off-flavours" have been identified and characterised. This allows their masking by the addition of flavours and provides a firm basis for their removal by processing. So it may be stated that the knowledge obtained in this project provides tools to better meet consumer preferences with respect to taste of pea-based NPFs.

With respect to texture, much knowledge has been obtained about a process that is of prime importance in texture formation, i.e. heat-induced gelling behaviour of pea proteins. This behaviour is affected by the protein composition and by at least one processing parameter (the rate of cooling). So varying the protein composition by selecting appropriate cultivars and varying processing parameters offer possibilities to produce a range of textures. This provides tools to food technologists to direct the texture of pea-based NPFs towards consumer demands.

However, no NPFs have been made and hence the texture and taste of pea based NPFs was not evaluated by sensory analysis and/or consumer tests. Furthermore, data concerning the properties desired by consumers and/or manufacturers are not yet available. Therefore, the question whether it is possible to produce NPFs that better meet consumer demands with respect to texture and taste cannot be answered unequivocally yet.

Taking the above answers together, it can be expected that by varying the composition of pea proteins, NPFs can be produced that cover a range of textures and tastes. Breeding tools to change the protein composition as well as a model to optimise the primary production of peas with respect to yield and environmental load are available in principle. Furthermore – albeit not investigated in the PROFETAS programme – processing techniques and conditions as well as the choice of ingredients besides proteins (fats, carbohydrates) are other techniques to adjust texture and taste of the NPFs. So it is to be expected that it is technologically feasible to produce NPFs having a broad range in taste and texture. However, whether within this range NPFs are present that better align with consumer wishes and preferences than present products, is still an open question.

A question that is related to technological feasibility is the following. Is it possible to design an NPF chain that delivers viable NPFs to the market? If we take viable products as products that meet the three P's (Planet, People and Profit), this more or less boils down to three questions regarding the NPF supply chain: Is it possible to design a chain that meets requirements with respect to environmental load, consumers demands and cost price? A methodology for chain design that takes into account these requirements separately has been developed. The methodology to integrate all three requirements is currently investigated. Apart from the methodologies, the project on chain design (Section 3.7) also offers some other insights:

- It was found that in terms of exergy the NPF supply chain is only slightly more efficient than the pork chain. This is mainly due to the fact that most of the exergy does not go into the NPFs but into the waste-stream caused by the non-protein fraction. Therefore, utilisation of this waste stream will result in a much higher efficiency of the NPF chain in terms of exergy. Furthermore, exergy analysis is only one way of looking at the environmental load of the chain. The results of other approaches (see Chapter 2) also clearly indicate a substantial reduction of the environmental load by changing from meat consumption to that of NPFs.
- Another finding was that initial cost calculations showed that the cost price of NPF could be less than or at least comparable to that of pork meat. Of course, these are merely indications, because neither the actual products nor the way and the scale on which they are going to be produced are known.

The production of NPFs will inevitably release non-protein fractions. Useful applications of these fractions are indispensable to enable a sustainable transition from meat protein to plant protein. Because of their economic value, food applications for these fractions are generally preferred over the use as feed, industrial raw material or energy source. Consequently, it was argued in Section 3.7 that oil crops (such as soy and rapeseed) are presently preferred over starch-rich crops such as peas. In addition, arguments such as protein yield per hectare and – in case NPFs are successful – changes in plant oil availability favour this conclusion. However, not only non-protein fractions are of importance for the decision which crops are to be used for NPF production. First of all, the quality of the NPFs has to be taken into account. In addition, numerous other factors are involved; they are being dealt with in Sections 6.2 and 6.3.

In the PROFETAS programme, pea proteins were taken as the model starting material to produce NPFs. Therefore, another relevant question for the PROFETAS programme as a whole is: Is the knowledge that has been generated within the technological projects also applicable to other plant proteins? The crop growth model and the methodology for chain design are generic and can be used for other plant proteins. The same holds for the insights obtained in options for the non-protein fractions. The technique developed for genetically modifying plants may also be useful for other legumes. The knowledge that has been generated in the two processing projects is more specific as is illustrated by the finding of Section 3.3: comparing the heat-induced gelling behaviour of the proteins from two legumes, soy and peas, showed similarities as well as differences. This leads to the conclusion that most of the knowledge obtained is applicable to other plant proteins. However, in particular in the processing part of the NPF supply chain the individual character of the proteins shows up, making generalisations more difficult.

Summarizing, as is evident from this section and the concluding paragraphs of the preceding sections in this chapter, much new knowledge and many new methods and models have been generated that contribute to NPFs production by industry. Furthermore, it is to be expected that a range of products can be made differing widely in product characteristics. However, it is not yet known whether NPFs can be made that better meet consumers wishes and preferences than the products currently available. The question whether or not a shift in the diet from meat to plant proteins is technologically feasible cannot be answered unequivocally, therefore.

REFERENCES

Andriot, I., Marin, I., Feron, G., Relkin, P., and Guichard, E. (1999), "Binding of benzaldehyde by β-lactoglobulin, by static-headspace and high performance liquid chromatography in different physico-chemical conditions", *Le Lait*, Vol. 79, pp. 577-586.

Apaiah, R.K., and Hendrix, E.M.T. (2005), "Design of a supply chain network for pea-based novel protein foods", *Journal of Food Engineering*, Vol. 70, pp. 383-391.

Apaiah, R.K., Hendrix, E.M.T., Meerdink, G., and Linnemann, A.R. (2005a), "Qualitative methodology for efficient food chain design", *Trends in Food Science and Technology*, Vol. 16, pp. 204-214.

Apaiah, R.K., Linnemann, A.R., and Van der Kooi, H. (2005b), "Exergy analysis: a tool to study the sustainability of food supply chains", *Food Research International (in press)*.

Beamon, B.M. (1998), "Supply chain design and analysis: Models and methods", *International Journal of Production Economics*, Vol. 55, pp. 281-294.

Bean, S.J., Gooding, P.S., Mullineaux, P.M., and Davies, D.R. (1997), "A simple system for pea transformation", *Plant Cell Reports*, Vol. 16, pp. 513-519.

Bishnoi, S., and Khetarpaul, N. (1994), "Saponin content and trypsin inhibitor of pea cultivars: Effect of domestic processing and cooking methods", *Journal of Food Science and Technology*, Vol. 31, pp. 73-76.

Chung, S., and Villota, R. (1989), "Binding of alcohols by soy protein in aqueous solutions", *Journal of Food Science*, Vol. 54, pp. 1604-1606.

Cousin, R. (1997), "Pea (*Pisum sativum* L.)", *Field Crops Research*, Vol. 53, pp. 111-130.

Daveby, Y.D., Aman, P., Betz, J.M., and Musser, S.M. (1998), "Effect of storage and extraction on the ratio of soya saponin I to 2,3-Dihydro-2,5-dihydroxy-6-methyl-4-pyrone-conjugated soya saponin I in dehulled peas (*Pisum sativum L.*)", *Journal of the Science of Food and Agriculture*, Vol. 78, pp. 141-146.

Daveby, Y.D., Aman, P., Betz, J.M., Musser, S.M., and Obermeyer, W.R. (1997), "The variation in content and changes during development of soya saponin I in dehulled Swedish peas (*Pisum sativum L.*)", *Journal of the Science of Food and Agriculture*, Vol. 73, pp. 391-395.

Davies, J., and Lightowler, H. (1998), "Plant-based alternatives to meat", *Nutrition and Food Science*, Vol. 2/3, pp. 90-94.

Dumont, J.P. (1985), "Diacetyl retention by pea protein used as an emulsifier", *Progress in Flavour Research - Developments in Food Science*, Vol. 10, pp. 501-504.

Dumont, J.P., and Land, D.G. (1986), "Binding of diacetyl by pea proteins", *Journal of Agricultural and Environmental Ethics*, Vol. 34, pp. 1041-1045.

Goudriaan, J., and Van Laar, H.H. (1994), *Modelling potential crop growth processes*, Kluwer Academic Publishers, Dordrecht, The Netherlands.

Grant, J.E., Cooper, P.A., McAra, A.E., and Frew, T.J. (1995), "Transformation of peas (*Pisum sativum* L.) using immature cotyledons", *Plant Cell Reports*, Vol. 15, pp. 254-258.

Gueguen, J. (1983), "Legume seed protein extraction, processing, and end product characteristics", *Qualitas Plantarum - Plant Foods for Human Nutrition*, Vol. 32, pp. 267-303.

Gueguen, J. (2000), "Pea proteins: New and promising protein ingredients", *Industrial Proteins*, Vol. 8 No. 1, pp. 6-8.

Heng, L., Van Koningsveld, G.A., Gruppen, H., Van Boekel, M.A.J.S., Vincken, J.P., Roozen, J.P., and Voragen, A.G.P. (2004), "Protein-flavour interactions in relation to the development of novel proteins foods", *Trends in Food Science and Technology*, Vol. 15, pp. 217-224.

Heng, L., Vincken, J.P., Hoppe, K., Van Koningsveld, G.A., Decroos, K., Gruppen, H., Van Boekel, M.A.J.S., and Voragen, A.G.P. (2005), "Stability of pea DDMP saponin and the mechanism of its decomposition", *Food Chemistry (submitted)*.

Ikedo, S., Shimoyamada, M., and Watanabe, K. (1996), "Interaction between bovine serum albumin and saponin as studied by heat stability and protease digestion", *Journal of Agricultural and Environmental Ethics*, Vol. 44, pp. 792-795.

Jeuffroy, M.H., and Ney, B. (1997), "Crop physiology and productivity", *Field Crops Research*, Vol. 53, pp. 3-16.

Kotas, T.J. (1995), *Exergy method of thermal plant analysis*, Krieger Publishing Company, Melbourne, Florida, USA.

Kropff, M.J., Bouma, J., and Jones, J.W. (2001), "Systems approaches for the design of sustainable agro-ecosystems", *Agricultural Systems*, Vol. 70, pp. 369-393.

Kudou, S., Tonomura, M., Tsukamoto, C., Shimoyamada, M., Uchida, T., and Okubo, K. (1992), "Isolation and structural elucidation of the major genuine soybean saponin", *Bioscience, Biotechnology and Biochemistry*, Vol. 56, pp. 142-143.

O'Kane, F.E. (2004), *Molecular characterisation and heat-induced gelation of pea vicilin and legumin*, PhD thesis, Wageningen University.

O'Kane, F.E., Happe, R.P., Vereijken, J.M., Gruppen, H., and Van Boekel, M.A.J.S. (2004a), "Characterisation of pea vicilin. 1. Denoting convicilin as the α-subunit of the Pisum vicilin family", *Journal of Agricultural and Environmental Ethics*, Vol. 52, pp. 3141-3148.

O'Kane, F.E., Happe, R.P., Vereijken, J.M., Gruppen, H., and Van Boekel, M.A.J.S. (2004b), "Characterisation of pea vicilin. 2. Consequences of subunit heterogeneity on heat-induced gelation behaviour", *Journal of Agricultural and Environmental Ethics*, Vol. 52, pp. 3149-3154.

O'Kane, F.E., Happe, R.P., Vereijken, J.M., Gruppen, H., and Van Boekel, M.A.J.S. (2004c), "Heat-induced gelation of pea legumin: A comparison with soybean glycinin", *Journal of Agricultural and Environmental Ethics*, Vol. 52, pp. 5071-5078.

O'Kane, F.E., Happe, R.P., Vereijken, J.M., Gruppen, H., and Van Boekel, M.A.J.S. (2005), "Gelation behavior of protein isolates extracted from five cultivars of *Pisum sativum* L.", *Journal of Food Science*, Vol. 70, pp. 132-137.

O'Keefe, S.F., Resurreccion, A.P., Wilson, L.A., and Murphy, P.A. (1991a), "Temperature effect on binding of volatile flavour compounds to soy protein in aqueous model systems", *Journal of Food Science*, Vol. 56, pp. 802-806.

O'Keefe, S.F., Wilson, L.A., Resurreccion, A.P., and Murphy, P.A. (1991b), "Determination of the binding of hexanal to soy glycinin and β-conglycinin in an aqueous model system using the headspace technique", *Journal of Agricultural and Environmental Ethics*, Vol. 39, pp. 1022-1028.

Owusu-Ansah, Y.J., and McCurdy, S.M. (1991), "Pea proteins: A review of chemistry, technology of production and utilisation", *Food Reviews International*, Vol. 7 No. 1, pp. 103-134.

Potter, S.M., Flores, J.R., Pollack, J., Lone, T.A., and Jimenez, M.D.B. (1993), "Protein-saponin interaction and its influence on blood lipids", *Journal of Agricultural and Environmental Ethics*, Vol. 41, pp. 1287-1291.

Price, K.R., and Fenwick, G.R. (1984), "Soya saponin I, a compound possessing undesirable taste characteristics isolated from dried pea (*Pisum sativum* L.)", *Journal of the Science of Food and Agriculture*, Vol. 35, pp. 887-892.

Price, K.R., Griffiths, N.M., Curl, C.L., and Fenwick, G.R. (1985), "Undesirable sensory properties of the dried pea (*Pisum sativum*). The role of saponins", *Food Chemistry*, Vol. 17, pp. 105-115.

Relkin, P., Mollé, D., and Marin, I. (2001), "Non-covalent binding of benzaldehyde to beta-lactoglobulin: characterisation by spectrofluorimetry and electrospray ionisation mass-spectrometry", *Journal of Dairy Research*, Vol. 68, pp. 151-155.

Sadler, M.J. (2004), "Meat alternatives – market developments and health benefits", *Trends in Food Science and Technology*, Vol. 15, pp. 250-260.

Schroeder, H.E., Schotz, A.H., Wardley-Richardson, T., Spencer, D., and Higgins, T.J.V. (1993), "Transformation and regeneration of two cultivars of pea (*Pisum sativum* L.)", *Plant Physiology*, Vol. 101, pp. 751-757.

Shan, Z., Raemakers, K., Tzitzikas, E.N., Ma, Z., and Visser, R.G.F. (2005a), "Development of a highly efficient, repetitive system of organogenesis in soybean (*Glycine max* (L.) Merr.)", *(in preparation)*.

Shan, Z., Raemakers, K., Tzitzikas, E.N., Ma, Z., and Visser, R.G.F. (2005b), "Effect of TDZ on pea plant regeneration from mature zygotic embryos in pea (*Pisum sativum*)", *(in preparation)*.

Shimoyamada, M., Ikedo, S., Ootsubo, R., and Watanabe, K. (1998), "Effects of soybean saponins on chymotryptic hydrolyses of soybean proteins", *Journal of Agricultural and Environmental Ethics*, Vol. 46, pp. 4793-4794.

Shimoyamada, M., Ootsubo, R., Naruse, T., and Watanabe, K. (2000), "Effects of soybean saponin on protease hydrolyses of β-lactoglobulin and α-lactalbumin", *Bioscience, Biotechnology and Biochemistry*, Vol. 64 No. 4, pp. 891-893.

Slinkard, A.E., Bhatty, R.S., Drew, B.N., and Morrall, R.A.A. (1990), "Dry pea and lentil as new crops in Saskatchewan: A case study", in Janick, J., and Simon, J.E. (Eds.), *Advances in new crops*, Timber Press, Portland, Oregon, USA, pp. 159-163.

Tijskens, L.M.M., Koster, A.C., and Jonker, J.M.E. (2001), "Concepts of chain management and chain optimization", in Tijskens, L.M.M., Hertog, M.L.A.T.M., and Nicolai, B.M. (Eds.), *Food process modeling*, Woodhead Publishing Limited, Cambridge, UK, pp. 448-469.

Tzitzikas, E.N., Bergervoet, M., Raemakers, K., Vincken, J.P., Van Lammeren, A., and Visser, R.G.F. (2004), "Regeneration of Pea (*Pisum sativum* L.) by a cyclic organogenic system", *Plant Cell Reports*, Vol. 23, pp. 453-460.

Tzitzikas, E.N., Bergervoet, M., Shan, Z., Raemakers, K., Vincken, J.P., and Visser, R.G.F. (2005a), "Transformation of pea (*Pisum sativum* L.) using a new regeneration system by a cyclic organogenic system", *(in preparation)*.

Tzitzikas, E.N., Vincken, J.P., De Groot, J., Gruppen, H., and Visser, R.G.F. (2005b), "Genetic variation in pea (*Pisum sativum* L.) seed globulin composition and their in planta processing", *(in preparation)*.

Van der Lans, I.A. (2001), *The image and the evaluation of fresh meat and meat substitutes in 200 and 2001 (Het imago en de evaluatie van vers vlees en vleesvervangers in 2000 en 2001, in Dutch)*, Product Boards for Livestock, Meat and Eggs, Rijswijk, The Netherlands.

Van der Vorst, J.G.A.J. (2000), *Effective food supply chains: Generating, modelling and evaluating supply chain scenarios*, PhD thesis, Wageningen University.

Vose, J.R. (1980), "Production and functionality of starches and protein isolates from legume seeds", *Cereal Chemistry*, Vol. 57, pp. 406-410.

Weaver, P., Jansen, L., Van Grootveld, G., Van Spiegel, E., and Vergragt, P. (2000), "Constructive technology assessment: The case of novel protein foods", in Weaver, P., Jansen, L., Van Grootveld, G., Van Spiegel, E., and Vergragt, P. (Eds.), *Sustainable Technology Development*, Greenleaf Publishing Ltd, Sheffield (UK), pp. 119-150.

Yin, X., Goudriaan, J., Lantinga, E.A., Vos, J., and Spiertz, J.H.J. (2003a), "A flexible sigmoid function of determinate growth", *Annals of Botany*, Vol. 91, pp. 361-371.

Yin, X., Lantinga, E.A., Schapendonck, A.H.C.M., and Zhong, X. (2003b), "Some quantitative relationships between leaf area index and canopy nitrogen content and distribution", *Annals of Botany*, Vol. 91, pp. 893-903.

Yin, X., and Van Laar, H.H. (2005), *Crop systems dynamics: An ecophysiological model of genotype-by-environment interactions,* (in preparation), Wageningen Academic Publishers, Wageningen.

Yin, X., Van Oijen, M., and Schapendonck, A.H.C.M. (2004), "Extension of a biochemical model for the generalised stoichiometry of electron transport limited C3 photosynthesis", *Plant, Cell and Environment*, Vol. 27, pp. 1211-1222.

Yin, X., Verhagen, J., Jongschaap, R., and Schapendonck, A.H.C.M. (2001), *A model to simulate responses of the crop-soil systems in relation to environmental change,* Note 129, Plant Research International, Wageningen.

CHAPTER 4

SOCIAL DESIRABILITY: CONSUMER ASPECTS

Joop de Boer, Annet Hoek
& Hanneke Elzerman

4.1 INTRODUCTION TO CONSUMER BEHAVIOUR[1]

Chapter 4 is the first of two chapters on the question whether a diet shift is socially desirable. It takes a behavioural perspective, whereas Chapter 5 is oriented towards processes at the level of organisations and markets. Accordingly, the main theme of the present chapter is how a diet shift is related to the behaviour of producers and consumers. The degree to which a shift "fits" into existing behavioural patterns is an important argument for its desirability. The same applies even more strongly to its future fit into the behavioural patterns of the next decades. Alternatively, whether a lack of fit will create an insurmountable problem depends on the feasibility of the measures that can be taken to mitigate the main shortcomings of the options.

What conditions make it attractive for producers and consumers to select NPFs instead of meat protein foods? From the perspective of producers, the answer may seem simple. The reasons for a producer to launch a new food product may be quite diverse, but they can always be translated into traditional business criteria, aimed at short-term and long-term profits. Given the consumer-oriented food market of today (Warde, 1997), it can simply be argued that the decisions of consumers will determine whether the introduction of a new food product will become a success or a failure. However, this statement is far too strong. Notably, producers are able to shape the food choices of consumers step by step in a certain direction (e.g. the direction of processed foods). Also, it is up to producers to develop new products and to decide whether they are ripe for the market. In sum, their role should not be neglected.

This chapter will focus on the factors that influence food choices of consumers. The analysis includes both short-term and long-term influences and it will pay due attention to the many linkages between the activities of

---
[1] Joop de Boer

food producers and consumers. These linkages are typical of the way in which food supply has been organised in modern society. They demonstrate that producers and consumers are almost continuously engaged in the exchange of signals about food-related opportunities and preferences (e.g. special offers, quality products). This does not mean, however, that the signals that they receive from each other are always clear. Consumers who buy meat products, for example, might have very mixed feelings about meat, but this attitude does not reveal what kind of alternative they would prefer. In other words, consumers' preferences are far from fixed and cannot entirely be read off their current purchases.

To get more insight into the various influences on food choices, this chapter argues that human behaviour is a very flexible phenomenon and that each particular manifestation of it can be the result of many determinants (De Boer, 2004). These determinants can be sorted into a logical order on the basis of the time frames that they involve. The fact that, for example, impulse buying has another time frame than consciously buying says something about the different underlying processes. Generally, the time frames that are relevant for behaviour range from short-term (i.e. taking less than a second) to long-term (i.e. taking almost a lifetime) and extremely long-term (i.e. taking many human generations).

The most obvious influences on behaviour are the perceptual and rational processes that enable a person to make sense of day-to-day events, such as an invitation to try some of the food. Because sense making is in essence a backward looking activity (Weick, 1995), the short-term influences on behaviour may refer to the taste of the food and the person's ideas about its origin. The behaviour of a person who is in doubt about the quality of a food product served by a host may be partly influenced by the bonds of convention and the fear of what this host will think (i.e. social processes). Also relevant may be the person's experiences with the business practices that are common in a certain food supply chain (e.g. fast food restaurants). In short, the behaviour in question is not only a function of processes within the person, but also of social and organizational processes that act as "proximal" causes of behaviour. These processes will often take days to decades and may change in a certain direction during the person's lifetime.

Moving from processes that are internal and proximal to more distal processes (i.e. long-term causes), we can see determinants of behaviour that will not dramatically change during the lifetime of an individual. These relatively stable processes can influence the person tasting the food, if, for example, he or she is drawn to beliefs about purity and danger that result from broadly shared worldviews (e.g. philosophies of life, beliefs about magical powers). These worldviews have gradually changed over the past millennium, due to a process of cultural modernization. Unlike mediaeval men and women, modern people will not expect solutions from magical powers, but they may still be sensitive to some of these beliefs under conditions of uncertainty.

A final category involves evolutionary processes, which have shaped human capabilities to cope with the environment, for example the ability to make a quick distinction between sweet (i.e. rich in calories) and bitter tasting (i.e. possibly poisonous) foods. The brain systems responsible for evaluating stimuli display a so-called "negativity bias", which means that negative stimuli (e.g. a suspect bitter taste) have a greater impact on information processing than do positive stimuli.

Figure 4-1 shows how the processes mentioned above can be arranged in a cascade-like framework. An important practical message of the framework is that the distal factors provide the context in which the more proximal or internal factors can have their effect. For example, the taste of a food may only be pleasurable for those persons who have already learned to appreciate the corresponding cuisine. Similarly, a "free-range" label will only have a moral effect on people who value animal welfare. The asymmetrical impact of positive and negative influences is evident from the following:

- The pleasure of eating might easily be spoiled by unpleasant ideas about the origin of the food.
- However, the unpleasant taste of a food will not easily be improved by pleasant ideas about its origin.

The framework of Figure 4-1 can help to generate information on the chances that a diet shift will be promoted or inhibited. Specifically, it is important to know whether all the influences on a particular behaviour point in the same direction and support the relevant changes. In the case that a particular behaviour is difficult to change, such as overeating, it is essential to combine as many influences as possible (e.g. organizational, social, rational and perceptual). It should be kept in mind, however, that each type of influence has its own time frame. For example, it will take more time to improve the social status of novel products than to increase the practical knowledge of consumers.

Accordingly, the framework opens the way to look at influences on behaviour from various perspectives. This will be done in the next sections, which describe three consumer-oriented research projects. Section 4.2 takes a long-term view on behaviour; it starts at the level of distal processes and analyses the socio-cultural changes in society that can make a diet shift more attractive or less attractive to producers and consumers.

In contrast, Sections 4.3 and 4.4 take a short-term view on behaviour; these projects start at the level of perceptual and rational processes to analyse consumers' reactions to novel products. More specifically, Section 4.3 addresses the way in which NPFs may replace meat in current dietary patterns. In addition, Section 4.4 tries to get more insight into the appreciation of NPFs, their appropriateness for various meals, and the way sensory preferences can be translated into product characteristics and physical parameters.

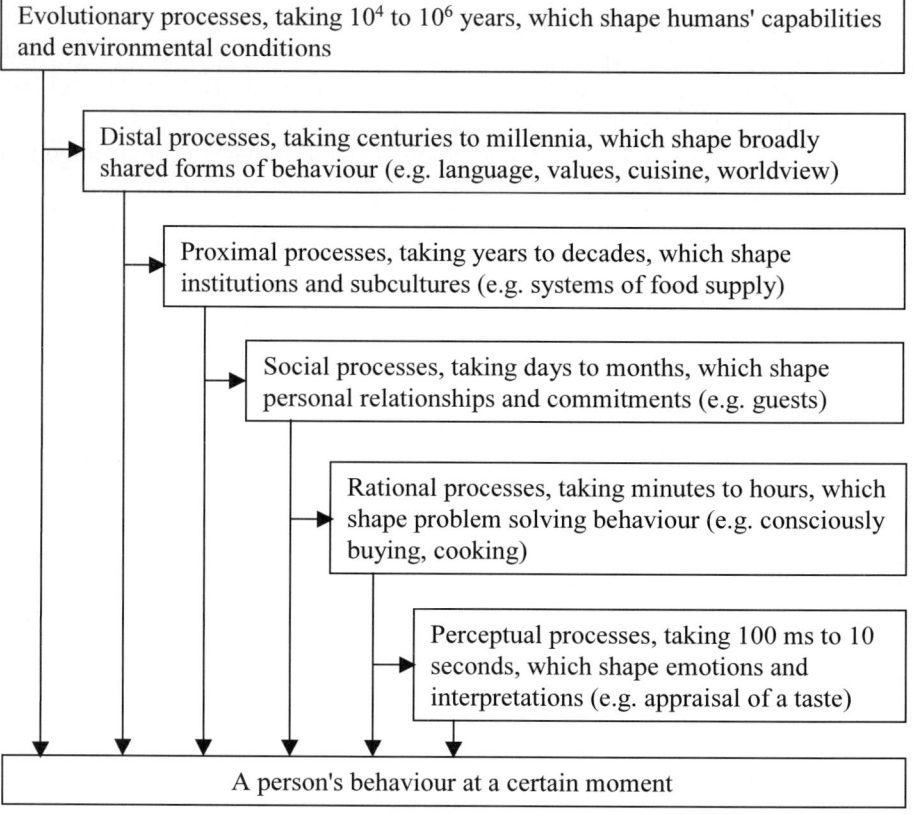

Figure 4-1. A cascade-like framework of influences on behaviour.

4.2 SOCIO-CULTURAL POTENTIAL[2]

4.2.1 Introduction

What is the potential for a diet shift in relation to long-term socio-cultural changes? In answering this question, this section focuses on food-related themes and lifestyles, which can provide incentives (but also disincentives) for consumers and producers to become less dependent on meat proteins. Some relevant examples are the increasing significance that Western consumers attribute to animal welfare and the growing appreciation of vegetarian meals, not only by consumers, but also by nutritionists. Given the various manifestations of these changes, the question should be raised whether they will continue to grow and make a substantial impact on the consumption of meat. An obvious alternative is that they will fade away like other food fads and fashions.

From a methodological point of view, studying the links between food choice criteria and long-term socio-cultural development is a challenging project. The main strategy chosen here is the development of a framework that sorts influences on behaviour into a logical order (Figure 4-1). Generally, a long-term development will create opportunities for food choices that match its general direction, whereas it will put constraints on others. Accordingly, it may be expected that those food choice criteria that appear to be part of a long-term change will have more impact in the future than criteria that are only based on short-term trends.

The combination of long-term and short-term approaches is not a simple task, as there are no databases and tools to support this type of research. Nevertheless, it is possible to use insights from the relevant disciplines (i.e. psychology, sociology, anthropology, history) as elementary building blocks, and to test at least some implications by small-scale "experiments" such as research on different versions of a questionnaire with built-in suggestions that unobtrusively remind consumers of meat's animal origin (Hoogland *et al.*, 2005; see below). In the next section, the role of Western modernization processes is analysed as an example of long-term influences on food choice

---

[2] Joop de Boer

criteria. These results will be combined with a brief description of the changes that have influenced food supply in the last few decades.

4.2.2 Results

Many of the links between current food choices and long-term socio-cultural development can be explained in relation to Western modernization processes. To put it simply, modern society can be distinguished from its predecessor by the potential democratisation of both its wealth (including meat eating) and its political process (Levine, 2001: 11). In terms of important periods in world history, the "modern" period is said to have started in 1900 and its predecessor in 1350, when Europe had to cope with the social, economic and political effects of the Black Death (Goldstone, 2005). In this "pre-modern" period, people could easily burst out into emotional behaviour, including violence against other people and animals. Their way of life was strongly dependent on their social and moral rank in society. It is against this background that a number of socio-cultural changes should be mentioned that are part and parcel to the overall process of modernization (see Table 4-1). They comprise:
- the increasing self-control considered typical of Western civilized man, such as the self-control of animal-like behaviour (since about 1500, see Elias, 1978),
- the rise of consumerism (or the belief that it is good to buy and use a lot of goods) among the middle classes (since about 1700, see Stearns, 2001), and
- the growing importance of an "engineering culture" characterized by the systematic application of scientific knowledge to societal issues (since about 1800, see Carroll-Burke, 2001).

To a certain extent the changes were supported by the mainstream of society, but they were also criticized by one or more counter-movements. The nineteenth century, for example, saw on the one hand a "democratization of meat" among European working-class families, influenced by the agricultural and industrial revolutions (Knapp, 1997). On the other hand, there were growing moral objections to the subjugation of animals, resulting in the foundation of the first vegetarian societies (Thomas, 1983).

Table 4-1. Main characteristics of three socio-cultural processes that mediate Western modernization.

Long-term changes	Direction of mainstream	Direction of counter-movement
Increasing self-control to weaken the link between impulses and behaviour (since about 1500).	Development of more predictable and civilized behaviour, suppressing every activity felt to be "animal", such as spitting or gobbling or the tendency to sniff at food.	Discovery by the upper and middle classes of "pacified nature" as an escape from civilizing rules, a source of pleasure and knowledge (reason for protests against the cruel treatment of animals).
Break with the rule that people should consume according to their rank in society (since about 1700).	Development of social arrangements (e.g. shops) and personal lifestyles (e.g. those of shoppers) in pursuit of the belief that it is good to buy and use a lot of goods.	From its early beginning in Britain, France, the Low Countries and parts of Germany and Italy, consumerism has provoked opposition, inspired by various moral, esthetical and political themes.
Break with the link between what is morally right and scientifically true (since about 1800).	Development of "engineering cultures", which use the powers of "engine science" in the laboratory for other cultural forms such as agriculture and medicine.	Rise of various subcultures concerned, among other things, with natural foods and holistic medicines, trusting the self-healing capacity of the human body.

The process of modernization brought many changes in dietary choice and culinary technique. Based on reports on the history of food, Table 4-2 summarizes a number of relevant differences between on the one hand the sixteenth/seventeenth century and on the other hand the beginning of the twenty-first century. Due to the prevailing prominent position of the court society in France, most of the changes were at first part of a French-style modernization before they became accepted more generally. As far as meat is concerned, many changes were particularly related to its animal origin. An interesting example is the practice of bringing the whole dead animal, or large parts of it, to the table, where the meat was to be carved by the master of the house or by distinguished guests. Research by the historian Flandrin (1999) shows that there were dozens of animal species served on the tables of the French aristocrats, although this number decreased between 1500 and 1650. There was, for example, a decreasing consumption of various large birds (e.g. swan). By way of contrast, the status of beef rose and much attention was paid to the particular cut of meat.

The list of observations in Table 4-2 suggests a number of significant diet shifts. Each of the differences must have had one or more proximal causes that explain how changes were created. For example, that members of the elite ate so many types of animal indicates that 17th century diet was still largely determined by fluctuations in availability as a consequence of the prevailing meteorological conditions (Flandrin, 1999). Interestingly, beef was considered "crude" and dismissed as indigestible by chefs in the aristocratic kitchens. Members of the elite left "gross" meats as well as most vegetables to the common people, whose stomachs were supposedly more robust. The elite ate only "delicate" fowl, relatively "light" fish and soft wheat bread.

Table 4-2. Some meat-related practices that have changed between the 16th-17th century (see Flandrin, 1999) and the beginning of the 21st century.

16th-17th century	Beginning of the 21st century
There were very large differences between high and low members of society.	A large part of the population of Western countries can afford to eat meat.
Rich people ate many types of animals, including various birds such as swans.	Consumers mainly choose a few types of animal.
Their cooks served large parts of the animal, which were carved at the table.	Consumers seldom serve whole animals; instead they serve cuts of meat.
It was a matter of good manners that upper class men should be able to cut meat from a pheasant still decorated with its feathers.	The cuts are bought at stores in which the carcasses have been hidden from the customer's eye.
Scientists agreed that the rich needed to eat birds to keep their intelligence and sensibility more alert.	The nutritional literature begins to appreciate the value of low-meat diets and vegetarian diets.
The working classes were considered best off eating large amounts of vegetables.	Nutritionists see vegetables as an essential part of each diet.
Local authorities tried to ensure an adequate supply of "good and honest" food.	Ensuring the provision of "good and honest" food is a task for supra-national authorities.

In the course of the 17th century, progress in the arts of butchery and cooking made it possible that the status of beef rose and that more attention was paid to the particular cut of meat. Accordingly, the serving of large parts of the animal to be carved at the table slowly went out of use. This decreasing practice is also connected with the gradual reduction in the size of the household and the transference of household activities to specialists (Elias, 1978).

Although the direct causes and the precise timing of these changes may not always be clear, their consequences are part of the long-term process of Western modernization. For example, people got fewer reminders that the meat dish has something to do with the killing of an animal. The practice of slaughtering has more and more been moved behind the scenes of social life (Vialles, 1994). According to Elias (1978: 120), this shift means that the mediaeval standard of feeling by which the sight and carving of a dead animal on the table were actually pleasurable, or at least not at all unpleasant, has been replaced by another standard by which reminders that the meat dish has something to do with the killing of an animal are avoided. Although this development is not uniform, the general direction of the changes seems to be the same. In many of our meat dishes the animal form is so concealed and changed by the art of its preparation and carving that during a meal one is scarcely reminded of its origin.

The long-term processes mentioned above can be complemented by processes that have taken place during the last few decades (see Table 4-3). These processes include shifts in the way people manage to organize their household, taking due account of differences in economy of scale. For example, if the costs of preparing a meal are compared per unit time of the eaters, a decreasing number of persons per household will make convenience food more attractive (Beardsworth and Keil, 1997; Warde, 1997). Other important processes refer to the way producers manage to supply foods and the way the authorities manage to control the public dimensions of food.

Table 4-3. Main characteristics of three proximal changes that have influenced food supply in the past decades.

Proximal changes	Main moderators	Direction of consequences
Shifts in the way people manage to organize their household.	Decreasing household size, less time spent on household activities, more income per person.	More demand for convenient products and ready-made meals, more tolerance of diverging food preferences within a household.
Shifts in the way producers manage to supply foods.	Growing influence of world markets, more emphasis on processing and packaging, less emphasis on primary production, supply chains more dominated by branded manufacturers and large retailers.	Growing number of products, more differentiation of qualities (taste, nutrition, health, convenience, moral concerns), more diverse points of sale (supermarkets, food courts, takeaways).
Shifts in the way the authorities manage to control the public dimensions of food.	Growing influence of supranational institutions, more emphasis on standardization, greater role for science-based notions of nutrition, health and animal welfare.	More communication about risk factors focusing on single nutrients (e.g. fatty acid profile) or functional ingredients.

One of the almost unnoticed consequences common to the shifts mentioned in Table 4-3 is their match with the long-term process of paying less attention to the meat-producing animal as a whole. Modern consumers seldom serve whole animals, but they serve cuts of meat that they have bought in stores in which the carcasses have been hidden from the customer's eye. Moreover, partly as a result of concerns about risk factors, such as saturated fatty acids, there has been a shift in consumption toward poultry and fish and away from beef and pork. As opposed to whole roasts, many consumers use products processed further, such as fillets.

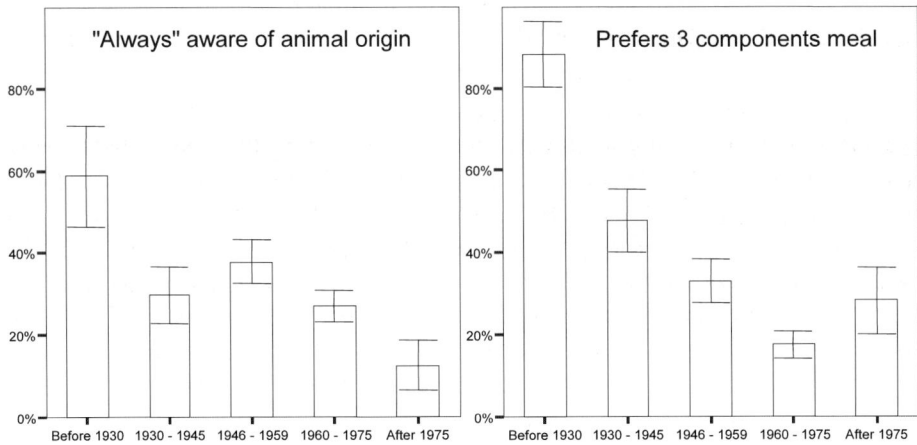

Figure 4-2. Share of consumers "always" giving thought to the animal origin of meat (left) and preferring a three components meal (right) in relation to year of birth. (Note: the bars are shown with standard errors; the sample of 313 supermarket customers is described in Hoogland et al., 2005.)

The psychological and socio-cultural implications of this development have not yet been fully explored. However, some results of research on consumers give an interesting clue. In May 2003, a sample of customers of a discounter plus one of a more expensive supermarket in the city of Rotterdam were asked to fill out a questionnaire, which contained items measuring food choice criteria and attitudes towards the association between meat and animals, such as the degree to which they give thought to meat's animal origin (see further details in Hoogland *et al.*, 2005). Only those customers were included who said that they ever bought meat. The results suggest that many consumers – at least those who live in a city – are not constantly aware of the animal origin of meat and that this awareness strongly decreases among the younger generations (left-hand part of Figure 4-2). Under the assumption that these differences between consumer generations reflect cultural changes on a time scale of decades, this result is in agreement with the long-term trend. Another interesting result is that the "three components" meal (meat, potatoes, vegetables) that was dominant in the Netherlands during the second part of the 20th century has lost significant popularity among the younger generations (right-hand part of Figure 4-2). This may indicate that meat is less used as the central part of the meal. Although it should be emphasized that these consumers had certainly not become vegetarians, their attitude towards meat's origin showed a remarkable sensitivity. After an unobtrusive suggestion that reminded them of meat's animal origin, they gave more weight to animal welfare as a food choice criterion than without that reminder.

4.2.3 Conclusions

Any attempt to summarize long-term processes in a few paragraphs should arouse suspicion, as it incurs the risk of taking the phenomena being described out of their historical context. Combining long-term and short-term influences on behaviour may easily give the impression of juxtaposing different types of work as if there is no difference between them and they can just be "added together". There is, however, no alternative to find out more about the potential for a diet shift in relation to long-term socio-cultural changes. Moreover, even if the direct causes and the precise timing of the various changes are not always clear, the general direction of their consequences is quite understandable.

The overall picture is that current manifestations of a certain ambivalence towards meat fit in a long-term process in which reminders of meat's animal origin have disappeared. This is a development that will continue to deepen in the future. The fact that many people are less aware of the animal origin of meat may be interpreted in terms of indifference toward the origins of proteins. This opens possibilities for NPFs, particularly in view of the decreasing popularity of the "three components meal" with its prominent cut of meat. If meat is less used as the central part of a meal, it will become feasible to design ready-made meals that contain more plant proteins and less meat or no meat at all (meat-free meals). If these meals are being developed and prepared by food producers, and consumers can choose such a meal without thinking about the source of the proteins, this strategy may even create a substantial shift from meat to plant protein foods without much consumer involvement.

However, although such a low-involvement approach will fit in a long-term socio-cultural development, it may not be the optimal strategy to pursue more sustainable food choices. One of its drawbacks is that it will reinforce mindless acceptance of technological changes. This mindless attitude is not in agreement with the preferences of consumers who have some affinity with one or more critical movements in society. These consumers want to be mindful of any potential value conflicts that technological innovations may bring about, including those associated with novel protein foods. Although they are only a small minority, their influence in society should not be underestimated. Therefore, it is of vital importance to involve both mainstream and critical consumers in discussions on food production methods.

Another reason to adopt a more transparent approach is related to the fact that meat will remain on the menu. The point is that people who are no longer aware of meat's animal origin will also be less inclined to pay attention to animal welfare. This process may have serious repercussions for other attempts to stimulate sustainable agriculture, for example, by promoting high quality meat from well-treated animals or by encouraging a high-plant and low-meat "Mediterranean" type of diet. From a sustainability perspective, these alternatives can go together with attempts to develop meat substitutes. Generally, an increase in the transparency of the food chain is likely to enhance sustainable food choices by producers and consumers.

4.3 SUBSTITUTION OF MEAT BY NPFs: FACTORS IN CONSUMER CHOICE[3]

4.3.1 Introduction

What can be done to replace meat in current dietary patterns? In the Netherlands, meat is still an important part of the meal. The market share of meat as a hot meal component is 76% (PVE, 2003), and a vast majority (>80%) consumes meat at dinner more than three days a week (Aurelia!, 2002). As noted in Chapter 3, meat-free plant-based products that are intended to replace meat, so-called "meat substitutes", were introduced in Europe during the last decades (Davies and Lightowler, 1998; McIlveen et al., 1999). However, the market for meat substitutes is still very small: about 1% of the total market for meat and meat products in the Netherlands (Aurelia!, 2002). This implies that in order to be successful, NPFs should be distinctive from meat substitute products currently on the market. To achieve a considerable reduction in the consumption of meat, NPFs should also be competitive with meat products (i.e. be better or cheaper). To develop such NPFs more information is necessary on consumer factors that play a role in the replacement of meat by meat substitute products, covering the whole consumption chain from product identification to repeated consumption over time.

This section is meant as a guide to NPF product development using the knowledge and the tools of food choice research. In studying factors that influence the choice for certain foods, three main components are usually

[3] Annet Hoek

distinguished: the Food, the Person, and the Environment (Shepherd, 1989). This section focuses on the Person, or the consumer, and the interaction with the Food, or product. Since pea-derived NPFs are not on the market yet, meat substitutes that are currently available were used as a case study. After a description of different consumer segments that are buying meat or meat substitutes, the role of product identification, the consumption experience, and repeated consumption will be discussed in relation to the overall acceptance of meat substitute products.

4.3.2 Results

Consumer segments

As part of the marketing process, specific target markets should be selected for NPFs. Therefore, the market has to be divided into groups of buyers with different needs, characteristics or behaviour, who might require separate products or marketing mixes. This is called market segmentation (Kotler *et al.*, 1999). Two studies illustrate the different consumer segments with respect to meat substitute products. Since NPFs will not be aimed at vegetarians primarily, non-vegetarian consumers of meat substitute products are considered particularly interesting.

For the first study (Hoek *et al.*, 2004), we used a representative sample of consumers (i.e. the Dutch National Food Consumption Survey 1997/1998). Non-vegetarian consumers of meat substitutes were compared to vegetarians and meat consumers with respect to socio-demographic and attitudinal variables. Both vegetarians (n=63) and non-vegetarian consumers of meat substitutes (n=39) were comparable for socio-demographic characteristics: higher educated, higher social class, living more in urbanised regions and smaller households than meat consumers (n=4313). Attitudes to food were assessed by the food-related lifestyle questionnaire (Grunert *et al.*, 1997), which is intended to assess attitudes with respect to ways of shopping, quality aspects, cooking methods, consumption situations and purchasing motives. We found that vegetarians (n=32) had more positive attitudes towards importance of product information, speciality shops, health, novelty, ecological products, social events, and social relationships than meat consumers (n=1638). The health consciousness scale (Schifferstein and Oude Ophuis, 1998) that was used to assess attitudes to health supported earlier findings that vegetarians are more preoccupied with health. However, food-related lifestyle and health attitudes of non-vegetarian meat substitute consumers (n=17) were much more in line with those of meat consumers.

In 2003, a survey was performed to collect new consumer data after the occurrence of several meat crises after 1998 (e.g. BSE and foot-and-mouth disease) and the resulting growth in the meat substitute market. It was aimed to provide more insight into factors and barriers acting on several levels of substitution of meat by meat substitute products. In this second study we used the usage frequency of meat substitutes as a basis for segmentation. Data was collected in the UK (a mature meat substitute market) and the Netherlands (a developing market) by means of a questionnaire that assessed demographic characteristics, food neophobia (the tendency to avoid new foods) by the Food Neophobia Scale of Pliner and Hobden (1992), food choice motives by the enhanced Food Choice Questionnaire (Lindeman and Väänänen, 2000; Steptoe et al., 1995), opinions on meat substitutes, and the desired similarity of meat substitutes to meat. The respondents (UK: n=235, 10% vegetarian; NL: n=318, 6% vegetarian) were classified into three categories: non-users (UK-45%, NL-69%), light/medium users (UK-35%, NL-16%) and heavy users (UK-20%, NL-15%). Among heavy users (meat substitute consumption at least once a week), the percentage of respondents that said they never eat meat was 17% in the UK and 29% in the Netherlands. We found no significant difference in overall food neophobia levels between the UK (mean food neophobia score 28.8) and the Netherlands (29.1). When user groups were compared within countries, it was found that non-users were more food neophobic than light/medium users (see Figure 4-3). However, again heavy users display a higher tendency to avoid unfamiliar foods compared to light/medium users, which might be explained by a particular lifestyle and values attached to food.

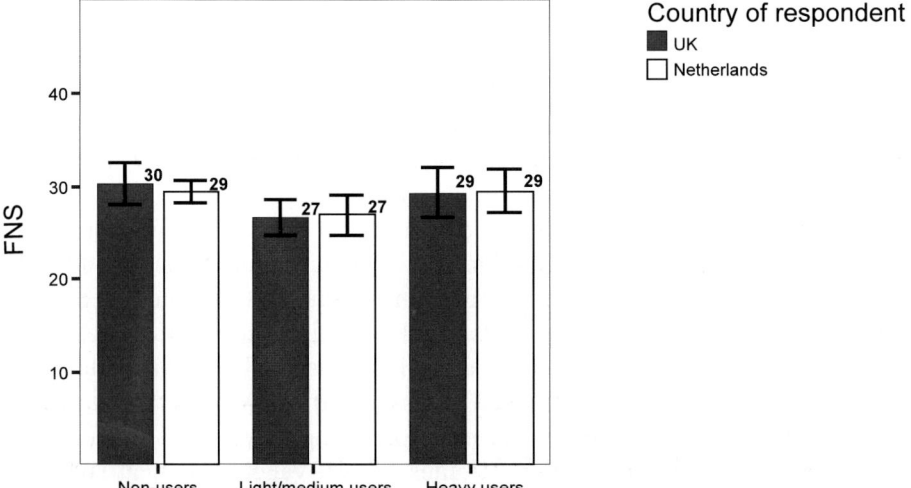

Figure 4-3. Food Neophobia Scores (FNS) of non-users, light/medium users (below once a week) and heavy users (once a week or over) of meat substitutes in the UK and the Netherlands. (The theoretical range of FNS is 10, very food neophilic, to 70, very food neophobic.)

With respect to overall differences in food choice motives between the two countries, Dutch respondents paid less attention to political values and more to price than respondents from the UK. Within countries it was found that a higher consumption of meat substitutes was related to higher importance attached to ethical food choice motives, such as ecological welfare (including animal welfare), political values, and natural content (preference for natural or organic food products). In contrast, non-users and light/medium users gave less weight to these motives. The non-users and light/medium users scored meat substitutes negatively for familiarity and luxury aspects and favourably for ethical aspects and weight control. As mentioned before, however, health and ethical aspects are not the main food choice motives of this group. In addition, respondents rated to which degree meat substitutes should resemble meat. Non-users and light/medium users indicated a preference for a meat substitute with a high similarity to meat for smell, texture, taste and appearance.

The outcome of both studies suggests that in order to attract new consumers, the focus should not be on health and ecological aspects of meat substitute products. Higher acceptance levels of meat substitutes in the UK might be explained by a greater number of vegetarians in that country and a higher interest in ethical aspects, which are both less pronounced in the Netherlands. When targeting non-users or light/medium users, more attention should be paid to luxury aspects and meat-like sensory properties. Unfamiliarity with the product and, to some extent, food neophobia can be a barrier to acceptance of NPFs.

Identification of substitutes

Obviously, the overall aim of PROFETAS implies that NPFs should be recognized as products that can be used instead of meat. Since it is not possible yet to develop an exact imitation of meat, it was assumed that NPFs should not mimic a meat product. The advantage of this assumption is that consumers do not expect a meaty taste or texture, which will reduce the risk of disappointment. On the other hand, "familiarity" plays an important role in acceptance (see above). Research indicates that consumers use categories to identify objects (Rosch and Mervis, 1975). Categories can be formed on the basis of perceived similarity and resemblance of products and are used to identify substitutes in the same or similar category. In view of this, a qualitative consumer study was carried out to explore which product attributes consumers used to identify a substitute for meat.

Since mainly extrinsic product characteristics are a source of information to consumers during shopping, the focus in this study was on the following product characteristics: information on label or package, package appearance, product appearance, and position in the supermarket. Semi-structured in-depth interviews were held with 15 respondents (students and employees of a non-agricultural university), in which they were asked to imagine themselves in an unfamiliar supermarket abroad looking for a substitute for meat, such as a vegetarian schnitzel. Almost every respondent answered that looking for the meat section in the store would be the first action. Subsequently respondents mentioned that they would pay attention to shape of the product, the product name ("vegetarian schnitzel"), the colour of the package ("green"), and phrases on the label such as "vegetarian", "meat substitute" or "soy". Results were also confirmed in a questionnaire-based survey (n=63). Thus, the position in the supermarket, product name and label information (reference to meat) can be crucial in the identification of NPFs as a substitute for meat in a meal. Future studies will further explore the role of associations (such as "green") and categorisation ("meat section") in identification of meat substitutes.

Consumption experience

Two main factors related to food choice are the physiological effects of foods and the sensory perception of physico-chemical properties of foods. These factors were studied by determining the satiating properties and changes in acceptability of existing meat substitutes and meat products after repeated ingestion, in relation to their sensory properties.

Satiety

Satiety is one of the physiological consequences of ingesting foods; it has been defined as the state that occurs after an eating episode, and that inhibits further eating (Blundell and Rogers, 1991). In the survey described above it appeared that respondents gave low scores for satiating properties of meat substitute products. In addition, an inventory of the nutritive value of products currently on the market revealed that some meat substitute products have a substantially lower protein content than meat, which is known to influence satiety sensation. Therefore, a study was performed to explore satiety scores of several meat substitutes compared to meat. Non-vegetarian students (n=28, 7 males) joined a consumption experiment during six days. Each day, meat or a meat substitute was randomly provided to each participant (males: 250 g; females: 200 g) in a lunch setting, and satiety measures (satiety scores on a 100 mm anchored line scale and amount eaten during a test meal) were taken until 2.5 hours after lunch. The selected products were four meat substitutes products and two meat products that were comparable in energy contents, but variable in protein contents. This resulted in an intake ranging from 20 g protein (two different meat substitutes with low protein content) to 71 g protein (two different meat substitutes with a high protein content) during those six lunches (for men).

The protein contents of the meat products selected for this study were in between these values. Comparing the subjective satiety sensations after consumption of these products, a trend was observed towards high protein meat substitutes resulting in lower hunger scores (mean hunger score is 33.2) than low protein meat substitutes (mean hunger score is 39.0), which was in line with our expectations. These preliminary results indicate that there is an observable difference in satiety sensations after consumption of different meat and meat substitute products that vary in protein content.

Sensory and hedonic evaluation

In preparation of a repeated exposure test (see the next paragraph), a pilot study was carried out to assess the liking scores of consumers for different, currently available meat substitutes that are used as ingredients in a meal. We were also interested in the similarity to meat perceived by these panellists. The consumer panel consisted of 23 non-vegetarian participants (mean age 26 years, 70% females), ranging from low to heavy users of meat substitutes. Six meat substitutes (minced, pieces, strips and cubes of different brands) were selected for this study and one reference meat product (chicken breast pieces). Unlabeled samples were presented in random order and 100 mm unstructured line scales were used for ratings. Results showed that chicken breast was the most liked sample before (smell and appearance only) and after tasting and had the highest intention to use. The meat substitute based on mycoprotein (pieces) was the most preferred meat substitute product after tasting and had the highest similarity to meat scores (overall similarity, and similarity in taste). Only 57% of the participants identified this product as a meat substitute, others were uncertain or thought the sample was a meat product. These results indicate that meat-like properties of meat substitutes play an important role in acceptance of these products. The role of sensory properties is more extensively described in Section 4.4.

Repeated consumption

Consumers may change their opinions on a food product after repeated consumption of the same food product over longer periods of time (Schutz and Pilgrim, 1958; Siegel and Pilgrim, 1958). Certain products seem to become "boring" or disliked, the latter resulting in low repeated purchases (Zandstra *et al.*, 2004). Concerning meat and meat substitute products we noticed that "heavy users" of meat substitutes eat these products usually a few times per week, in contrast to meat consumers who eat meat during the hot meal five times per week, or even more. Are meat substitutes too boring after a while? To achieve a successful replacement of meat by NPFs in the long run, insight into product properties and other factors playing a role in long-term acceptance of these products is essential. As an example of how this can be investigated, we organized a consumer in-home use test with repeated exposure (twice a week during a period of ten weeks) to either a commercial meat substitute product or a reference meat product. The main outcome of such a study is the change in liking of meat substitute products compared to a reference meat product after repeated consumption in a realistic setting. By comparing several meat substitutes, it is possible to assess, for example, whether a meat substitute that shows more similarity to meat is more acceptable over time than a meat substitute with less meat-like properties.

4.3.3 Conclusions

The food choice studies described above illustrate the variety of factors that may play a role in the replacement of meat by meat substitutes. Various scientific disciplines, such as nutritional science, food science, psychology, and marketing should be consulted in further research. The main point of this section is that targeting new (non-vegetarian) consumers for NPFs may offer interesting opportunities to distinguish these products from current meat substitutes. Given the sheer number of these consumers, this strategy may ultimately have the most beneficial effects in terms of environmental sustainability. However, this segment of consumers does not share vegetarian ideologies and will not be attracted by the environmental argument. These consumers tend to choose more conventionally and seem to prefer a meat-like product, as was replicated in an actual tasting test. Meat-like characteristics (e.g. appearance, packaging) also have a role for the identification of NPFs as a substitute for meat. Product characteristics such as satiating properties (referring to relative protein content) need attention in product development of NPFs as well.

4.4 SUBSTITUTION OF MEAT BY NPFs: SENSORY PROPERTIES AND CONTEXTUAL FACTORS[4]

4.4.1 Introduction

Sensory characteristics play an important role in the acceptance of foods. In order to achieve a transition from meat consumption towards more

---

[4] Hanneke Elzerman

sustainable NPFs based on plants, these foods should have sensory characteristics that are appealing to consumers. A consumer-driven approach is seen as the key to success for new product development, as can be concluded from the large number of publications on this topic (Costa et al., 2001; Van Trijp and Steenkamp, 1998). This approach, in which consumer wishes are taken as a starting point for product development, is also used here. The goal of the present project is to explore methods that may identify consumers' sensory expectations and preferences of NPFs. These methods can be used as a toolbox for consumer-driven product development.

As mentioned in Chapter 1, it is not considered feasible to strive for NPFs that can replace large pieces of meat, such as steaks or cutlets. Therefore, we chose to focus on NPFs that can be used as ingredients in a dish (in minced form, slices or pieces). This implies that the whole dish, and not just the NPF-ingredients, will determine the acceptance of these foods. The appropriateness of the use of these NPF-ingredients in different dishes seems to be of crucial importance for the acceptance of NPFs. That is why appropriateness in the context of a dish has been a central topic of this project.

The role of the context of a dish or meal on the acceptance of foods has hardly been studied before. Turner and Collison (1988) studied the role of meal components (starter, entrée, sweet) on the acceptance of a meal. Stallberg-White and Pliner (1999) tested the hypothesis that the addition of familiar flavours to novel staple foods would decrease the "neophobia" (fear of new things) by consumers. They found that the addition of a familiar sauce to a novel food increased subjects' willingness to taste it. Context in the meaning of a situation in which a food is eaten has been the subject of several studies. The importance of other "contextual factors" and the appropriateness of the use of foods in a situation have been recognized, for example, by Cardello and Schutz (1996) Schutz (1994), Rozin and Tuorila (1993) and Meiselman et al. (2000). It was concluded that when, where, how, with whom or with what you eat a food are important determinants for the acceptance of foods.

The project focused on the acceptance of "meat substitute ingredients" by consumers in a naturalistic environment. Consumers tasted these products in a university dining hall, which is a far more normal setting for consumers than sensory booths. The products were evaluated both in several dishes and "as such". In addition, we looked deeper into the products with a trained sensory panel that described them with objective sensory attributes. Furthermore, we looked at meat substitutes in a broader context by analysing the differences in appropriateness between meat substitutes and meat (products) in several food use situations.

An important complicating factor is that consumers can only give sensory preferences after they have eaten or at least seen or smelled a product. Pea-based NPFs are new foods that do not exist yet. Therefore, commercially available meat substitutes have been used in the various studies. Meat substitutes were defined as products that have been developed to substitute meat in a dish. Fish, cheese, nuts, eggs, etc. are not considered meat substitutes.

4.4.2 Results

Consumers' experiences and expectations of meat substitutes

From a market exploration in 2001, we concluded that there are over 150 meat substitutes on the Dutch market. These products vary from plain tofu and vegetarian burgers and schnitzels to meat substitute ingredients (pieces, mince, slices), snacks and sandwich toppings. The main ingredient is mostly soy protein, but it can also be wheat protein, mycoprotein (fungi) or a mixture of vegetables. It was stated in an earlier report on NPFs that most available meat substitutes were not well accepted by consumers. The main deficiency would be the texture properties of most products (Sijtsma et al., 1996).

To get more insight into the factors that are important in the acceptance of meat substitutes, we conducted qualitative consumer research using focus group discussions (Greenbaum, 1998; Krueger and Casey, 1988). Experiences of 46 consumers who had some experience with meat substitutes were elaborated. Consumers discussed why, when and how they used meat substitutes, and what their opinion was on these products (both from a practical and from a sensory point of view). After the general part of the focus group, consumers discussed the appropriateness of the use of meat substitutes in different dishes that were shown on photographs. The focus group discussions were concluded with a small tasting session in which consumers tasted meat substitutes in a dish and discussed their liking for the products.

Many positive and negative aspects of meat substitutes were mentioned. The remarks that were made can be divided into general and sensory aspects that describe the products. General remarks were mainly on the image of meat substitutes. "Meat substitutes" was considered a bad name, there was concern for genetic modification, and meat substitutes were found unnecessary products. Also, the lack of information on the package (on the origin of ingredients, preparation, and recipes) was often mentioned. Health aspects were mentioned both in a positive and in a negative meaning (low fat, high protein were found positive aspects, whereas low protein and artificial flavourings were considered negative aspects). Many consumers believed that meat is needed by children, in particular for its vitamin and mineral content. Finally, meat substitutes were often called expensive.

Sensory remarks that were made about the appearance of meat substitutes were that most products looked like meat products. For some consumers this was a positive and for others this was a negative aspect. Also, some consumers liked the taste and the texture of meat substitutes, whereas others disliked it. Furthermore, negative flavour aspects that were mentioned included: bland taste or too spicy, chemical aftertaste, dryness, stickiness, softness, sponginess, hardness, compactness, and toughness. Positive remarks included: chicken-like texture, granular texture, crispy crust, and neutral taste.

Most consumers found the use of novel protein foods appropriate in the meals that were shown on the photos (soup, pasta, rice, wrap, salad meal, and pizza). However, some consumers rejected the use of meat substitutes on pizzas, in meal salads and in soups. The results of these focus group discussions were used as a basis for further studies.

Appropriateness and liking of meat substitutes based on visual information

Based on the focus group discussions and pilot studies, we hypothesized that the context of the dish influences the acceptance of meat substitutes. To get more insight into the role of appropriateness on the acceptance of meat substitutes, we developed an Internet questionnaire in which we could show photographs of many meat substitute meal combinations to many consumers in a relatively quick way. The main goal of this questionnaire was to find out whether consumers find meat substitutes appropriate in different dishes and what kind of meat substitutes would be the most (or the least) appropriate. The on-line questionnaire consisted of:
- General appropriateness questions (e.g. how appropriate do you find the use of a meat substitute in a soup?)
- More specific appropriateness questions, based on photographs of combinations of meat substitutes and various dishes (e.g. how appropriate do you find the use of this meat substitute in this soup?)
- Questions on sensory aspects of meat substitutes

The study yielded 251 completed questionnaires. The main results were that a pasta dish, a rice dish or a wrap (pancake with filling) were considered more appropriate for the use of meat substitutes than a soup, a pizza or a meal salad. These results are in line with those of the focus group discussions. The top 3 of the most appropriate and least appropriate combinations of meat substitutes and dishes are shown in Table 4-4. As can be concluded from the table, most consumers prefer "familiar" combinations in which a meat substitute looks like the meat product it is replacing. There were no large differences between different consumer groups.

Table 4-4. Top 3 of the most appropriate combinations of dishes and meat substitutes and of the least appropriate combinations (out of 30 combinations in total).

Most appropriate		Least appropriate	
Spaghetti	Mince	Wrap	Slices
Wrap	Mince	Rice	Slices
Rice	Pieces	Pizza	Cubes

Brown was considered the most appropriate colour for meat substitutes, as it was indicated to be a positive colour by 80% of the respondents. Green was the least appropriate colour (indicated as a positive colour for meat substitutes by only 15% of the respondents). Other favoured properties for meat substitutes were: soft, smooth, crispy, seasoned, spicy and meat-like flavour.

Situational appropriateness of meat substitutes

The focus group discussions and a pilot study suggested that consumers find the use of meat substitutes less appropriate than meat in certain situations. For example, when consumers want to eat something "luxurious", they find a meat substitute not appropriate, unlike several meat products. This is in line with the findings of the survey described in Section 4.3. To learn more about the situational appropriateness of meat substitutes we developed a questionnaire. The situations were based on focus group discussions and the products that were used in this questionnaire covered a large part of the meat substitute product types. Meat products featured in this questionnaire as well. The questionnaire showed a photograph of the product and a list of 22 situations. Respondents had to rate the appropriateness of the products in each of the situations on a 5-point scale (1=not at all appropriate and 5=very appropriate). This project is currently being carried out. The results can be used by marketers to better position meat substitutes.

Appropriateness and liking of meat substitutes based on consumption

Sensory consumer studies provided insight into important factors for the acceptability of meat substitutes. In this study, about 100 consumers tasted combinations of dishes with meat substitutes and meat substitutes as such. The meat substitutes and dishes that were used in this study are shown in Table 4-5. A standardized method for preparation and serving of the dishes was developed first. All dishes were served hot, except for the salad. The samples were consumed in a university dining room. The respondents scored the samples (in a randomised order) on appropriateness of the use of meat substitutes in dishes, and on expected liking, overall liking, liking of appearance, taste and texture. The data are currently being analysed.

Table 4-5. Meat substitutes (with their main ingredient) and dishes used in the sensory consumer study (25 combinations of meat substitutes and dishes were tasted). Products 1-5 were pieces and product 6 was mince.

Meat substitute	Dish
Product 1 (wheat protein)	Dish 1: Rice with curry sauce
Product 2 (tofu (soy))	Dish 2: Rice with sweet & sour sauce
Product 3 (mycoprotein)	Dish 3: Rice with peanut sauce
Product 4 (wheat protein)	Dish 4: Spaghetti with tomato sauce
Product 5 (soy protein)	Dish 5: Tomato-vegetable soup
Product 6 (mycoprotein)(mince)	Dish 6: Pasta salad

Sensory description of meat substitutes

To better understand consumer liking and to be able to give directions to product developers in terms of product specifications, we conducted a descriptive analysis of meat substitutes. A sensory panel with 18 panellists was trained for Quantitative Descriptive Analysis® of 12 commercially available meat substitutes (in pieces and minced) (Stone *et al.*, 1974). The pieces were described by 21 sensory attributes and the minced meat substitutes by 22 sensory attributes (appearance, smell, taste, and texture). The appearance of the products was similar in colour, but differed in the size of the granules. The composition of the products was quite different. Most products (such as tofu) contained soy protein as the main ingredient, whereas others were made of a mixture of soy protein, wheat protein, or pea protein. One product was made from mycoproteins (see Section 3.1). Figure 4-4 shows two spider plots of the scores of the minced meat substitutes on different texture and flavour attributes. Please note that the products in Figure 4-4 are different from the ones mentioned in Table 4-5 and that the plots are only meant to illustrate the type of results.

Preliminary results suggest that the panel detected large differences between the products, especially for sour and rye bread flavour, saltiness, bitterness, toughness and juiciness. These descriptive data can be related to the sensory consumer data to find out which product properties of meat substitute ingredients consumers like or do not like.

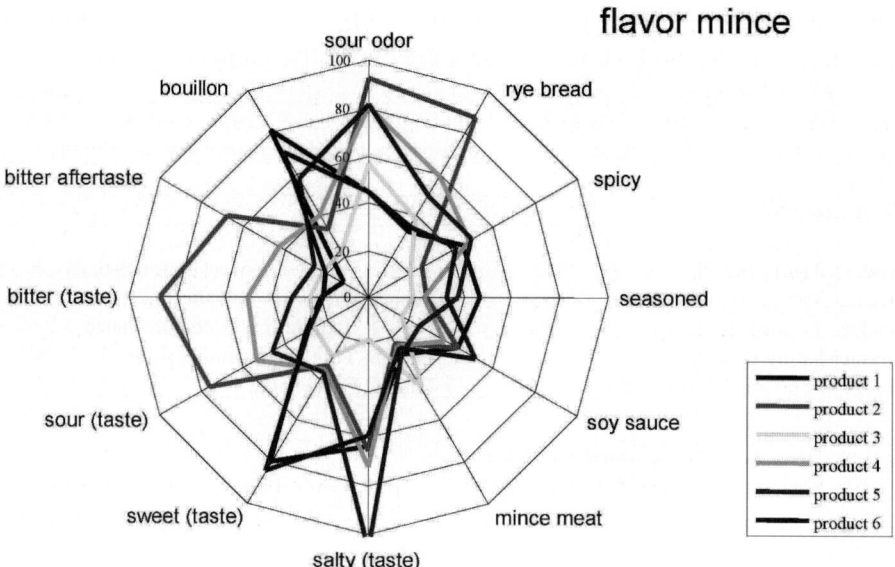

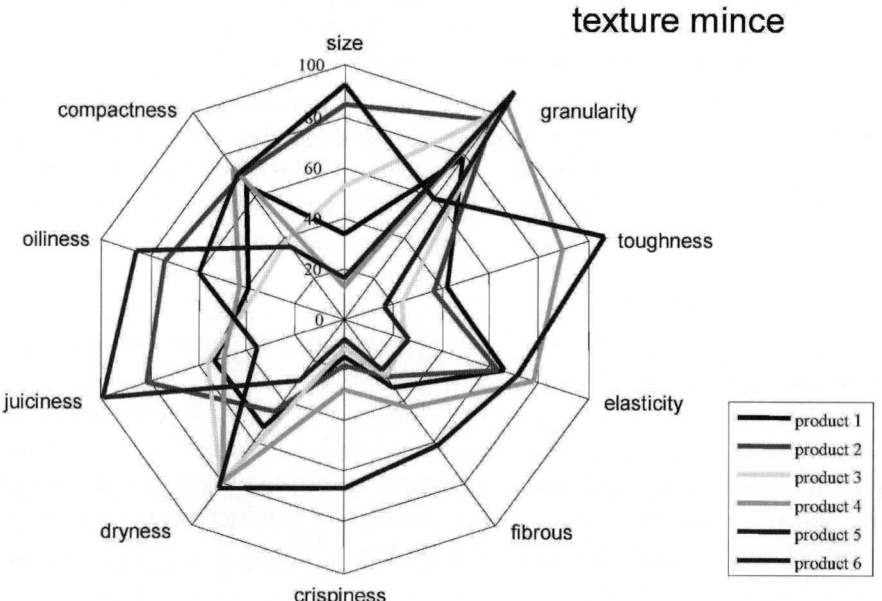

Figure 4-4. Spider web plots of the mean scores of the sensory panel on 22 sensory attributes (11 flavour and 11 texture attributes) for minced meat substitutes. The figure on top shows the flavour attributes (taste: bitter, sour, sweet, and salty; and odour: bouillon, sour, rye bread, spicy, seasoned, soy sauce, and minced meat), and the figure below shows the texture attributes (based on appearance: size; and based on mouth feel: granularity, toughness, elasticity, fibrous, crispiness, dryness, juiciness, oiliness, and compactness).

4.4.3 Conclusions

All studies taken together can give more insight into consumers' sensory preferences of meat substitutes and, more importantly, which sensory attributes and contextual factors are responsible for these preferences. In the focus groups and in the appropriateness questionnaire consumers gave their opinions about liking and appropriateness of meat substitutes based only on descriptions of the foods, food names and visual information. From these studies it can be concluded that only a small group of consumers is open to new products that are really different in appearance and flavour from the existing meat substitutes. However, the majority of consumers want meat substitutes to have a meat-like flavour and a brown colour. As far as the appropriateness of the use of meat substitutes in different meals is concerned, most consumers want combinations that are more or less familiar to them. They want the meat substitute to take exactly the same place in the dish as the meat it is replacing.

If these preferences are confirmed in other consumption studies, they can be coupled to the sensory data of the descriptive panel. Ultimately, these preferences could then be "translated" into measurable product properties, and used by product developers for the design of NPFs. However, valid instrumental (physical and chemical) methods for the measurements of all sensory characteristics of solid foods are not yet available (Rosenthal, 1999). In order to make this translation, evidently more research needs to be done in this field.

4.5   CONCLUSIONS[5]

This chapter was intended to examine whether a diet shift is socially desirable in view of the preferences of consumers and producers. The underlying notion was that the better a diet shift fits into the behavioural patterns of current and future generations, the more desirable it is. In addition, it was argued that a lack of fit does not have to cause an insurmountable problem if measures can be taken to mitigate the main shortcomings. Because these issues cannot be analysed directly, several indirect approaches were used. Based on a long-term view on behaviour, the potential for a diet shift in relation to socio-cultural changes was examined. At the more detailed level of food choices and sensory experiences

---

[5] Joop de Boer

consumer-directed methods were developed that may guide NPF product development.

One of the most salient results of this chapter is the contrast between, on the one hand, a series of impressive changes in dietary choices during the last few centuries and particularly during the last few decades and, on the other hand, the observation that an individual will not easily change his or her food preferences from one day to the next. This contrast underlines the value of our analytical framework, which sorts influences on behaviour into a logical order (Figure 4-1). Its cascade-like structure expresses the view that a long-term development will create opportunities for food choices that match its general direction, whereas it will put constraints on others. According to Section 4.2, there is a favourable socio-cultural context for decisions that make consumers and producers less dependent on meat proteins. However, it appears that the currently available meat substitutes will not become popular without additional measures. The consumer studies described in Sections 4.3 and 4.4 clearly showed that many consumers left alone with a choice between a currently available meat substitute and meat would prefer the latter.

The consumer studies also demonstrated that NPFs should be meat-like products that have the same place in the dish as meat. These results confirm the notion mentioned in Section 4.1 that people will habitually look for what is familiar when they are trying to make sense of something, such as an invitation to try a new product. This retrospective character of sense making can explain that non-vegetarian consumers keep relying on distinctions drawn in the past and that they evaluate meat substitutes by using meat-based criteria. Product developers should keep in mind that people are only able and willing to re-examine and revise their existing concepts at certain moments of change (i.e. discontinuously instead of continuously). Only clearly perceived benefits of a new product may stimulate them to supplement existing concepts and criteria. Importantly, the evidence presented in this chapter suggests that the ecological or moral benefits of NPFs will not be sufficient to change these consumers' minds.

By using currently available meat substitutes as a model, it was possible to develop several tools that may guide NPF product development, even though the most sensory and other characteristics of the former and the latter are likely to differ. A drawback of this approach may be that the current meat substitutes are in fact sold in a niche market and that they are almost twice as expensive as the cheapest meats. In contrast, pea-derived NPFs may be developed to produce cheap protein products for multiple purposes. Whether these products will be attractive and acceptable in terms of relevant consumer motives remains to be seen. This chapter has demonstrated, however, that there is now a robust set of tools to develop NPF products in a consumer-driven way.

REFERENCES

Aurelia! (2002), Market of meat substitutes in the Netherlands (Vleesvervangers in Nederland 2002, in Dutch), Aurelia!, Amersfoort, The Netherlands.
Beardsworth, A., and Keil, T. (1997), Sociology on the menu: An invitation to the study of food and society, Routledge, London.
Blundell, J.E., and Rogers, P.J. (1991), "Satiating power of food", Encyclopedia of Human Biology, Vol. 6, pp. 723-733.
Cardello, A.V., and Schutz, H.G. (1996), "Food appropriateness measures as an adjunct to consumer preference/acceptability evaluation", Food Quality and Preference, Vol. 7, pp. 239-249.
Carroll-Burke, P. (2001), "Tools, instruments and engines: Getting a handle on the specificity of engine science", Social Studies of Science, Vol. 31, pp. 593-625.
Costa, A.I.A., Dekker, M., and Jongen, W.M.F. (2001), "Quality function deployment in the food industry: A review", Trends in Food Science and Technology, Vol. 11, pp. 306-314.
Davies, J., and Lightowler, H. (1998), "Plant-based alternatives to meat", Nutrition and Food Science, Vol. 2/3, pp. 90-94.
De Boer, J. (2004), "A psychological view on industrial transformation and behaviour", in Olsthoorn, A.A., and Wieczorek, A.J. (Eds.), Sciences for industrial transformation: Views from different disciplines, Kluwer, Dordrecht, The Netherlands, pp. 27-45.
Elias, N. (1978), The civilizing process. I. The history of manners, Basil Blackwell, London.
Flandrin, J.L. (1999), "Dietary choices and culinary technique, 1500-1800", in Flandrin, J.L., Montanari, M., and Sonnenfeld, A. (Eds.), Food: A culinary history from antiquity to the present, Columbia University Press, New York, pp. 403-417.
Goldstone, J.A. (2005), "Efflorescences and economic growth in world history: Rethinking the "Rise of the West" and the Industrial Revolution", Journal of World History, Vol. 13, pp. 323-389.
Greenbaum, T.L. (1998), The practical handbook and guide to focus group research, Sage, Thousand Oaks, CA.
Grunert, K.G., Brunsø, K., and Bisp, S. (1997), "Food-related life-style: Development of a cross-culturally valid instrument for market surveillance", in Kahle, L., and Chiagouris, C. (Eds.), Values, lifestyles and psychographics, Erlbaum, Hillsdale, NJ, pp. 337-354.
Hoek, A.C., Luning, P.A., Stafleu, A., and De Graaf, C. (2004), "Food-related lifestyle and health attitudes of Dutch vegetarians, non-vegetarian consumers of meat substitutes, and meat consumers", Appetite, Vol. 42, pp. 265-272.
Hoogland, C.T., De Boer, J., and Boersema, J.J. (2005), "Transparency of the meat chain in the light of food culture and history", Appetite, Vol. 45, pp. 15-23.
Knapp, V.J. (1997), "The democratization of meat and protein in late eighteenth- and nineteenth-century Europe", Historian, pp. 541-551.
Kotler, P., Armstrong, G., Saunders, J., and Wong, V. (1999), Principles of marketing, second European edition, Prentice Hall Europe, London.
Krueger, R.A., and Casey, M.A. (1988), Focus groups: A practical guide for applied research, Sage, Thousand Oaks, CA.
Levine, D. (2001), At the dawn of modernity: Biology, culture, and material life in Europe after the year 1000, University of California Press, Berkeley, CA.

Lindeman, M., and Väänänen, M. (2000), "Measurement of ethical food choice motives", Appetite, Vol. 34, pp. 55-59.

McIlveen, H., Abraham, C., and Armstrong, G. (1999), "Meat avoidance and the role of replacers", Nutrition and Food Science, Vol. 99, pp. 29-36.

Meiselman, H.L., Johnson, J.L., Reeve, W., and Crouch, J.E. (2000), "Demonstrations of the influence of the eating environment on food acceptance", Appetite, Vol. 35, pp. 231-237.

Pliner, P., and Hobden, K. (1992), "Development of a scale to measure the trait of food neophobia in humans", Appetite, Vol. 19, pp. 105-120.

PVE (2003), Market research 2002: Meat, figures and trends (Marktverkenning 2002: Vlees, cijfers en trends, in Dutch), Product Boards for Livestock, Meat and Eggs, Zoetermeer, The Netherlands.

Rosch, E., and Mervis, C.B. (1975), "Family resemblances: Studies on the internal structure of categories", Cognitive Psychology, Vol. 14, pp. 573-603.

Rosenthal, A.J. (1999), Food texture: Measurement and perception, Aspen Publishers, Gaithersburg, ML.

Rozin, P., and Tuorila, H. (1993), "Simultaneous and temporal contextual influences on food acceptance", Food Quality and Preference, Vol. 4, pp. 11-20.

Schifferstein, H.N.J., and Oude Ophuis, P.A.M. (1998), "Health-related determinants of organic food consumption in the Netherlands", Food Quality and Preference, Vol. 9, pp. 119-133.

Schutz, H.G. (1994), "Appropriateness as a measure of the cognitive-contextual aspects of food acceptance", in MacFie, H.J.H., and Thomson, D.M.H. (Eds.), Chapman & Hall, Glasgow, UK, pp. 25-50.

Schutz, H.G., and Pilgrim, F.J. (1958), "A field study of monotony", Psychological Reports, Vol. 47, pp. 559-565.

Shepherd, R. (1989), "Factors influencing food preferences and choice", in Shepherd, R. (Ed.), Handbook of the psychophysiology of human eating, Wiley, Chichester, pp. 3-24.

Siegel, P.S., and Pilgrim, F.J. (1958), "The effect of monotony on the acceptance of food", American Journal of Psychology, Vol. 71, pp. 756-759.

Sijtsma, L., Rabenberg, M., Janssens, R., and Linnemann, A.R. (1996), Sensory aspects of Novel Protein Foods (Sensorische aspecten van Novel Protein Foods, in Dutch), DTO werkdocument VN13, Delft, The Netherlands.

Stallberg-White, C., and Pliner, P. (1999), "The effect of flavor principles on willingness to taste novel foods", Appetite, Vol. 33, pp. 209-221.

Stearns, P.N. (2001), Consumerism in world history: The global transformation of desire, Routledge, London.

Steptoe, A., Pollard, T., and Wardle, J. (1995), "Development of a measure of the motives underlying the selection of food: The food choice questionnaire", Appetite, Vol. 25, pp. 267-284.

Stone, H., Sidel, J., Oliver, S., Woolsey, A., and Singleton, R.C. (1974), "Sensory evaluation by quantitative descriptive analysis", Food Technology, Vol. 28, pp. 24-34.

Thomas, K. (1983), Man and the natural world, changing attitudes in England (1500-1800), Allan Lane/Penguin Books, Harmondsworth.

Turner, M., and Collison, R. (1988), "Consumer acceptance of meals and meal components", Food Quality and Preference, Vol. 1, pp. 21-24.

Van Trijp, J.C.M., and Steenkamp, J.E.B.M. (1998), "Consumer-oriented new product development: Principles and practice", in Jongen, W.M.F., and Meulenberg, M.T.G. (Eds.), Innovation of food production systems: Product quality and consumer acceptance, Wageningen Pers, Wageningen, pp. 37-66.

Vialles, N. (1994), Animal to edible, Cambridge University Press, Cambridge.

Warde, A. (1997), Consumption, food and taste: Culinary antinomies and commodity culture, SAGE, London, UK.
Weick, K.E. (1995), Sensemaking in organizations, Sage, Thousand Oaks, CA.
Zandstra, E.H., Weegels, M.F., Van Spronsen, A.A., and Klerk, M. (2004), "Scoring or boring? Predicting boredom through repeated in-home consumption", Food Quality and Preference, Vol. 15, pp. 549-557.

CHAPTER 5

SOCIAL DESIRABILITY: NATIONAL AND INTERNATIONAL CONTEXT

Joop de Boer, Marike Vijver,
Lia van Wesenbeeck, Claudia Herok
& Onno Kuik

5.1 INTRODUCTION TO SOCIETAL ASPECTS[1]

This is the second of two chapters on the question whether a diet shift is socially desirable. Whereas the preceding one took a behavioural perspective, the present chapter is oriented towards processes at the level of markets and institutions. In terms of Chapter 1, it should be repeated that there is a difference between factors that contribute to the opening of a "window of opportunity" and factors that increase the chances that the open window will be used in a particular way. The latter refers especially to the competition between old and new technologies. From the field of evolutionary economics it is known that the successful introduction of alternatives to an old technology will depend on the characteristics of the "selection environment" (Glynn, 2002) – a concept that does not just include market forces, but also institutional arrangements, such as procedures to get authorisation for novel foods. Therefore, the main theme of this chapter is how a diet shift may "fit" into the food-related political and economic developments of the next decades.

The actors who are making selection decisions can be divided into two main categories. The first category refers to the "proximate decision makers" (Lindblom, 2001) in industry and government who take the economic and political decisions that play a major role in the marketplace. The second category refers to all the members of society who participate in the market and in the political processes. Figure 5-1 sketches an extremely simplified picture of the long and heterogeneous chains of interactions that connect the preferences of members of society at one end to various performances at the other end. It shows that society's actual progress in the direction of sustainability is the result of decisions on (a) the type of products that are

---

[1] Joop de Boer

made available to consumers and (b) the methods of production that are allowed. The decisions of entrepreneurs and government officials on these topics are steered to a certain extent by their anticipation of the response of consumers in the market and of citizens in the political process.

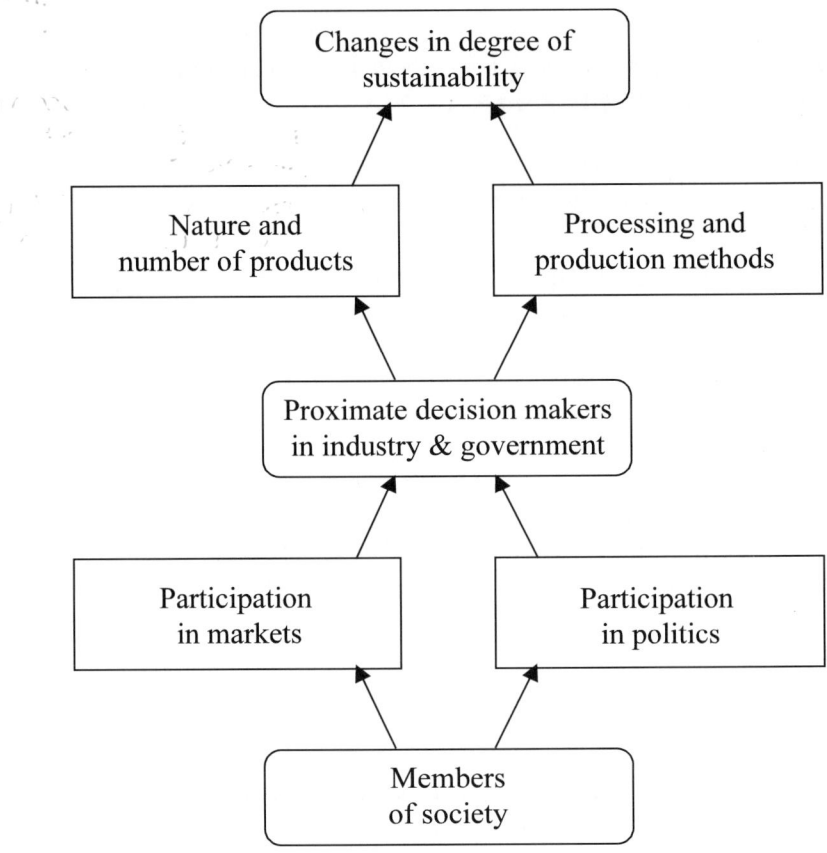

Figure 5-1. Impacts by members of society on changes in sustainability via markets and politics.

The notion of competing technologies suggests that there is an easy answer to the question whether a diet shift is desirable for society. Given the coupling between food, agriculture and sustainability, it might be said that society wants the most sustainable technology to win. This view has to be modified, however. First, it should be noted that the technology behind NPFs is at an early stage of development. Hence, competition is not possible yet. Second, competition according to the rules of the market system does not guarantee that the most sustainable technology will win. In addition to market forces, a broad range of political and economic factors is involved.

A key factor in this context is the nature of new technologies. Technology development is not a completely deliberate, planned and systematized activity, which is basically controlled by foresight. Instead, it is part of a world that is changing and malleable. Developing alternatives to an established technology may require many innovations and these will not occur automatically as the outcome of a linear process. For example, the recent history of GM tomato ketchup shows that many accidental factors have affected the content and the timing of decisions on this product – it took more than twenty years to develop the world's first genetically engineered whole food (i.e. tomato ketchup), but after an economic life of two years this product disappeared from the market as a result of trade policies (Harvey *et al.*, 2002).

It should be stressed at this point that the early stages of technology development can be extremely sensitive to the uncertainty surrounding the technical promise of innovations (Podolny and Stuart, 1995). In the case of NPFs, for example, it is still an open question which crops will be used and which by-products may be of interest (see Chapter 6). The various options will also have consequences for the stakeholders of other technologies and the supporters of other lifestyles. Because of such uncertainties, the development of an innovation can be seen as an unfolding network of very heterogeneous elements, which include connections with the scientific community, academia-industry interfaces, and wider social patterns of collaboration, collegiality and competition (Murray, 2004). In addition, there are many interactions with more general driving forces, such as agro-industrialization and globalisation.

Taking due account of these conditions of uncertainty, there are still opportunities for the initiators or stakeholders of a new technology to monitor and influence the selection environment in which their products have to survive. Such an approach should be made compatible with the notion that technology development takes place in a dynamic world. It requires advanced tools such as policy analysis and econometric modelling, as well as a skilled examination of the guidelines, regulations and laws of international institutions with regard to novel foods. Accordingly, the present chapter is intended to give a summary of what these tools can and cannot realize. One of the consequences of the uncertainties is that every approach, every project and every section has its own time perspective.

As a start, Section 5.2 analyses the ways in which government policies may affect food choices – not just directly, but primarily indirectly, in fact. This policy analysis can only be sensibly done in retrospect, based on a skilful reconstruction of the recent past. This study, focusing on the

Netherlands, specifies how food production and consumption have become parts of a commercial market, which allowed bulk supplies for lower prices. It was in this context that various pre-conceptions about animal feed (i.e. mainly grains, but also by-products) could shape political discussions on protein conversion factors when the notion of a "protein crisis" was launched in the late 1960s. Another salient example of the role of pre-conceptions is the market failure – as a meat substitute – of TVP, a processed protein product that was developed in the same period primarily from a technological point of view. Against this background, some contrasting visions of a future diet shift are explained.

Section 5.3 employs two types of econometric modelling to simulate macroeconomic developments for the period until 2020. Although the results are projections for the future, it should be noted that the models are based on economic transactions in the past. This becomes particularly clear with regard to the question whether meat prices will continue to decrease in the future, just like the low-priced bulk supplies mentioned above, or whether they will increase due to as yet unprecedented meat shortages in the world market. This contrast underlines the added value of employing complementary economic models.

Finally, Section 5.4 describes how the current rules of international institutions may affect the marketing of novel protein foods. This international institutional context is also particularly relevant for discussions on policy options, such as instating subsidies or levies, which may be considered by government officials and entrepreneurs as instruments to support novel technologies.

5.2   NATIONAL POLICIES AND POLITICS[2]

5.2.1   Introduction

Food involves a great variety of governmental policy issues. This study concentrates on the ways in which diverse policies have influenced the proportion of animal and plant proteins in our diets. At first sight, governments may seem to have little to do with personal food choices, plus most people would not appreciate policies meddling in their affairs. The ways in which policies affect our food choices are rather indirect and

---

[2] Marike Vijver

implicit, however. Probably most policies that have consequences for maintaining or shifting diets do not intend to influence people's diets at all. As a consequence, the normative, political aspect of protein politics is not primarily found in discussions between politicians, interest groups and citizens, but is more a side effect of decisions on other issues and the result of the production and consumption practice.

The important question here is how certain views and customs reached a point beyond discussion. Which actions and choices have made certain policy goals evident and discharged other solutions? The outcome of interactions in the practice of food production and consumption implicitly favours certain diets over others. What seems normal now, such as eating meat on a daily basis, or governmental intervention in an agricultural crisis, probably was not always normal. Therefore, this study delves into the process in which those events have been made normal. This process of normalisation (Brown, 1999) may show how certain views, roles, interests and activities became dominant over others.

Uncovering the indirect ways in which policies affect food requires a way of studying that does not start from the current practice of food production and consumption, but from the question how this current food practice came about. Politics, technology and markets are interconnected spheres of human society; politics, technology and markets exert influences on one another in a reciprocal way. This means that production and consumption are, at least partially, facilitated by a political "infrastructure", implicitly favouring certain products or production processes over others. As such, contemporary practice provides what we shall eat and what is easily available. For that reason, this study focuses on the political aspects of food production and consumption in the Netherlands.

5.2.2 Results

Although most of the developments described occurred in other Western-European countries also, this study focussed on the Netherlands. An ideal starting point was found to be a report published in 1847, in which the physician G.J. Mulder first brought up the issue of proteins and the societal effects of the lack of proteins in most people's diets. The subsequent analytical steps describe how developments in society, in the economy and in politics to some extent ordered food production and consumption by constructing the guiding principles for change, the spheres of activity (such as market and regulatory practices) and the roles of the actors involved. As an illustration, this section will focus on a later period, when the question whether people should eat less meat started to receive considerable public attention. Detailed references to the original Dutch sources can be found in Vijver (2005: 80-84).

Presenting a diet shift as a way to feed the world

During the late 1960s and early 1970s, experts expected that there would not be enough protein by the 1980s, considering the global population growth. In 1967, the Food and Agriculture Organisation of the United Nations spoke of "the daily increasing protein crisis" and recommended a series of measures to address it. In 1972 and 1973, a series of harvests failed, increasing the sense of concern regarding world food supply. Some Dutch interest groups, scientists, and political actors put forward that it might be better if the developed countries would consume more plant proteins and less animal proteins, leaving more food for people in developing countries. The notion received attention from the media. The second report to the "Club of Rome" (Mesaroviç and Pestel, 1974) was used for support, as it mentioned that 7 kilos of grains are required to deliver 1 kilo of meat. In a document prepared for the World Food Conference in Rome in 1974, the progressive Dutch political parties wrote:

> "In fact, a form of cannibalism still exists on this earth, be it in a modern, concealed form. It exists in the rich countries, where people feed themselves at the expense of people in the poor countries." (Translated from the original Dutch text.)

As one of the 29 goals in their recommendations for a policy dealing with food shortages, they suggested putting a brake on the consumption of meat in the Netherlands and Europe. Levying meat was considered an option, and good and active education by the government on correct and balanced eating patterns was considered necessary.

Seeking a diet shift through eating less meat

The prevalent quantities of meat production and consumption were questioned. The former minister of agriculture and former chairman of the European Commission, Sicco Mansholt, argued in 1974:

> "The question is for how long we can continue the conversion of 10 kg plant protein into 1 kg animal protein, in view of the need in developed countries, the scarcity and the increasing prices. In time, this leads also politically to an untenable situation." (Translated from the original Dutch text.)

Arguments for lowering meat consumption to feed more people were often linked to criticism of modern animal production. The conversion factor, the amount of edible plant material it would cost to deliver a unit of meat, was frequently treated as part of an opposition to intensive animal farming methods along with other issues: animal welfare, environment, human health (such as fat content and hormones in meat), the position of workers in production (such as the position of farmers in developing and developed countries) and certain philosophies of life.

In the meantime, producing and consuming animal products such as meat had become the normal thing to do. Within the production and consumption practice, producers, consumers and animal products all obtained certain prescribed roles and mutual interactions more or less fixed these roles. In the 1970s, meat production and processing kept increasing and the consumption of meat remained relatively stable. The notion of partly shifting diet, whether it seemed sensible or not, was not part of a routine, whereas the consumption and production of actual amounts of animal and plant products were part of everyday life. Meat and other animal protein products were, therefore, themselves important in refuting the notion of partially shifting from animal to plant products.

As far as the sales of animal products did not defy the notion of a shift towards plant products, representatives of animal food production, science and policy argued the value of such a shift on a number of grounds. One of these was the conversion factor. One author argued that 21 units of mainly edible plant material are needed to produce 1 unit of meat, but scientists researching feed and feed producers stated that it was much more favourable and could be 3 to 1. The percentage of grains in feed had been reduced from 62% in 1961 to 30% in 1972/1973. This reduction was mainly caused by high prices of grain in the European Union, the consequent imports of grain alternatives, the use of by-products and offal from the food processing industries and the technological competence of the feed industry.

It was also argued that eating less meat in developed countries would not necessarily lead to a greater availability of food in developing countries. For example, in 1975, Dutch minister of agriculture, Van der Stee, argued that a EU-policy aimed at lowering the meat consumption was not a realistic option. He mentioned as one of the reasons that eating less animal products in the EU would probably only lead to a lower production of grains by the United States, because the developing countries lacked the money to buy the grains instead.

No consensus was reached on the view that a diet shift would solve problems with the world food supply. Moreover, the solution in terms of "eating less" did not concur with the "eating more" guiding principle of the market. Hence, effort was put into refuting the argument that eating less meat would solve world hunger. As was the case with other discontents, the wider grievances with the prevalent meat production and consumption were dismissed.

Seeking a diet shift through producing more plant proteins

By the late 1960s, the isolation of plant proteins from material such as fungi, leaves and soy to make "meat alternatives" for human consumption showed potential profitability for food companies. Raw plant materials were relatively cheap, for example, soy-meal is a by-product of soy oil production. Also, the petroleum industry took an interest in developing so-called "single cell protein" by growing yeast on petroleum (Pyke, 1970).

Expectations of a world protein shortage and the urgent problems of hunger fostered more interest in "synthetic proteins". The United Nations stimulated a search for new protein sources, both for feed and for direct human consumption (via a "Task force on bioconversion of organic residues for rural communities"). Food scientists were trying to find efficient ways to produce proteins. Plant protein products derived through food technology were expected to be more efficient and cheaper than meat. These "synthetic protein foods" were intended for developed countries. The complex processing needed to produce these foods would tie the economic advantages to countries with a high consumption of animal proteins, where they could replace processed meat products. In the UK, mycoprotein was developed with financial support from the British government; later it formed the basis for Quorn^{TM}. In the late 1960s, a product named "Texturised Vegetable Protein" (TVP) was marketed in the Netherlands. TVP had been developed in the US and was made of soy protein supplemented with flavourings. Dutch consumers did not warm to TVP. Despite the high expectations of TVP as a promising food innovation, it failed on the market. As an observer noticed:

> "TVP dangled between art and artificial. Its name alone. I can picture it on the menu: TVP with salad!" (Translated from the original Dutch text.)

For food companies, however, there remained opportunities to make meat products with a little less meat and a little more plant, so-called "extending". The technology of making plant protein products did not go to waste and was applied in some processed meats or snacks. Meat was partly or completely replaced by texturised proteins and flavourings in order to decrease production costs and even to improve healthiness of the product. A barrier to be slighted was the European law on food that prohibited plant proteins in meat. But in the early 1980s, it became legal to mix soy proteins with meat proteins under certain conditions.

With regard to the consequences of the new products for human health, policymakers and nutrition experts were ambivalent; this can be understood in the light of earlier concerns about highly industrialised foods. On government request, in 1970 the Health Council of the Netherlands presented a number of detailed research guidelines to ascertain the safety of proteins. The council wrote further that they could not inform the government on the food situation in developing countries, because of a lack of information. The situation in the Netherlands, however, was considered such that protein shortages might occur among the young and the elderly; hence these groups could possibly benefit from "synthetic" proteins.

Nutritional scientists tested TVP on rats and since the growth of these test animals was retarded, it was recommended that TVP should not be fed to children, although it was unclear what the effects of TVP on children would be. The product seemed to be an excellent plant source of proteins and iron. In 1968, the provisional recommendation by the bureau of nutrition education was for adults not to eat TVP more than twice a week. Also, an official advised against feeding TVP to elderly people.

5.2.3   Conclusions

In the period analysed, a diet shift was considered in the form of eating less animal-based and more plant-based protein foods as a possible solution for the expected problems with the world food supply. The notion of a shift was put forward in two distinct ways. On the one hand, the notion was linked to criticism with regard to aspects of modern animal production, with the emphasis on eating less meat. On the other hand, the notion was linked to efforts to increase the profitability of food with the emphasis on producing new plant-based protein foods. Neither the "eat less meat", nor the "produce more plant-based alternatives" notions led to a diet shift, because no broad support was obtained for the view that a diet shift would be a solution for problems perceived with the world food supply. Although the notion to "produce more plant-based protein foods" was in line with the "provide more" guiding principle of the market ("consumerism") and also with the increasing role of scientific research on food development, producers failed to make the product desirable for consumers. Moreover, it did not match the worries about naturalness and health, which keep returning with the progress of industrialisation.

The analysis demonstrates – among other things – that a food practice and its guiding principles act as a framework for future developments. Shifting to NPFs may follow the prevalent trend of adding value to food through technological innovation and may fit the prevalent food practice in which the marketisation and industrialisation of food is essential. It also fits the recent trend to improve the competitiveness of the agro-industry by focussing on an environmentally friendly niche product. From the producers' point of view, NPFs are mainly a possibility to improve sales. For contemporary consumers, NPFs fits the "eat more" directives, as Nestle

(2002) calls them in her book on food politics. However, it is questionable whether NPFs can be promoted as a contribution to sustainable consumption patterns. The "eat more" option means basically that production and consumption will increase without a guarantee that environmental pressure will be reduced. The impact on sustainable protein production would be larger by simply eating less meat. Accordingly, if sustainability and long-term social acceptability is the goal of a diet shift, the "eating less meat" option should be considered a viable alternative, on par with replacing meat by producing more plant protein based alternatives. This requires that policymakers become more conscious of the assumptions underlying their actions to improve sustainability. In this way, they can try to create a more divergent food system with different approaches to food, so that "new plant protein based alternatives" will not necessarily become more normal than "eating less meat".

5.3   EUROPEAN AND GLOBAL ECONOMIC SHIFTS[3]

5.3.1   Introduction

This section is devoted to potential future macroeconomic developments during the period until 2020 in order to explore what a shift towards plant proteins might mean in the context of European and global development. The study focuses on implications of changes in protein production and consumption patterns for international trade and world agricultural markets, under different assumptions regarding economic, political and technical developments. The analysis builds on earlier work on the relationship between population growth, increasing incomes, the non-linear shape of the meat demand schedule, projected meat demand, feed requirements, and land use. It is a challenge for world markets in the decades to come to supply sufficient meat and, hence, feed grains and concentrates. However, there are different opinions on whether this is feasible.

In spite of continued growth of global population and consumption, projections by international agencies (Bruinsma, 2002; FAPRI, 2004; OECD, 2004) continue to indicate that with adequate effort it should be technically feasible to satisfy future demand, and that this will not cause major disruption of the agricultural markets (Delgado *et al.*, 1999).

[3] Lia van Wesenbeeck & Claudia Herok

However, it is argued (Keyzer *et al.*, 2005) that under an assumed growth in per capita income a significant underestimation of future demand for meat and cereal feeds will occur. Three mechanisms are essential to this argument. First, per capita demand for meat depends primarily on per capita income. Hence, the differences in per capita incomes *within* countries must be accounted for. Second, the Engel curve that links per capita consumption and income was shown to display a sigmoid relationship, as the poor segments of the population tend to abstain from meat consumption until their income reaches some lower threshold, while rich consumers become satiated beyond an upper threshold. In countries with a relatively high average demand for meat, still a large fraction of the population may just have started to consume meat. Third, the rise in meat demand requires an additional supply of feed that cannot be met by the traditional, pasture and crop residue based technologies, as meat demand rises faster than local crop production, while an increasing fraction of demand is for non-ruminants, and continued urbanization makes it more difficult to expand traditional animal husbandry with livestock roaming around the homestead.

In addition to an assessment of the consequences of a transition for agricultural production and trade, a second objective of this study is to contribute to the methodological development of Applied General Equilibrium (AGE) models in agriculture. Therefore, two models were used. On the one hand, the stylised General Equilibrium Model of Agricultural Trade (GEMAT) model is used, with relatively few commodities and activities and highly simplified empirical relationships. On the other hand, the Global Trade Analysis Project (GTAP) model was employed with its elaborate database with detailed information on production, consumption and trade flows for 66 regions and 57 products. The models are clearly complementary in their different foci and strengths. GTAP is based on a well-documented and constantly improved theoretical framework, which allows for a broad analysis of the economy. The GEMAT model, however, is a dedicated and practical representation of the main mechanisms relevant to the problems at hand. Though its coverage may be less encompassing, it is richer in specific detail. For example, it imposes physical constraints on the maximum attainable yield per ha for different crop types in different regions, explicitly incorporates constraints on feed composition from dietary requirements of ruminants and monogastrics, and models the production of many feed items as by-products of food production. Both models have been adapted and used for analysis of trade flows under different assumptions regarding economic, political and technical developments.

It has been anticipated that the two different models will give different outcomes when a transition from meat to NPFs is introduced. By tracing back the differences in model outcomes to differences in assumptions that underlie the models, both the models and the assumptions were screened and improved, thus providing additional information. By starting out from two different vantage points and working together in this way, it will be better possible to come to grips with the complicated mechanisms that are involved in transmitting the effects of a transition from meat to NPFs.

5.3.2 Results

GEMAT

First of all, a "business as usual" (BAU) scenario has been defined, which represents a situation where standard assumptions on consumption and feed technology are made, i.e. in this scenario no lifestyle changes and no restrictions on animal diets are included. This scenario is used as the benchmark against which the other scenarios are compared, and therefore, it is calibrated as well as possible to reflect actually observed patterns of production and consumption, including a low preference for vegetarian products, and a low preference for agricultural services at present. The goal of the numerical exercises with GEMAT described here is to illustrate the effects of alternative model structures, given a change in model parameters, in this case a change in preferences in the EU region from meat to novel protein foods. Given this objective, four alternatives are discussed:

1. SHIFT, where the only change made with respect to the BAU scenario is the change in preferences for meat and NPFs, realised by assuming a 20% decrease (in the CES share parameter) with respect to meat products and a 100% increase (in the CES share parameter) with respect to NPFs for the EU consumers of all lifestyles.
2. GREEN, where the change in preferences for meat is accompanied by an increasing preference for agricultural services, realised by increasing the CES share parameter for agricultural services by 50% for the EU consumers of all lifestyles.
3. DIET, which is the GREEN scenario complemented with the requirement that animal diets need to be respected.
4. LIFESTYLE, which is the DIET scenario in which lifestyle changes are incorporated.

The results show major differences between alternative assumptions on model structure, and adding the requirement that animals have certain dietary requirements that need to be respected, in particular, has an important impact on the scenario results. Secondly, it is noteworthy that the response of the model with respect to the changes in the model parameters defining the scenarios is rather "violent" in all but the DIET and LIFESTYLE scenarios. This is something which requires further research, and it derives from the lack of institutional structure that is imposed on the model, and from the fact that calibration of the model is far from completed, which causes the model to overstate the importance of agriculture in the economies of the regions, which in turn leads to the extremely large model responses. Therefore, it should be stressed that the results are illustrative only. Yet, it is interesting to note that the LIFESTYLE scenario clearly shows that the driving force of meat demand in developing countries is the shift in lifestyle that accompanies an increase in per capita income: in the low and middle income countries, meat demand continues to rise, but meat demand in the EU and also in the other high-income countries falls. In the EU, this is due to the assumed shift in preferences, accompanied by the lifestyle effect of satiation in meat at high income levels. In the other high-income regions, the effect is entirely caused by satiation effects, as more people shift from the intermediate to the rich, satiated lifestyle.

All cases, except the other (non-EU) high-income regions (under the DIET and LIFESTYLE scenarios only), display a decrease in the pressure on land, which presents the effects of a transition on the shadow prices of the land use bound, i.e. it indicates the loss associated with not taking more land into production than is currently allowed in the model (which guarantees that no nature land is taken into production). Therefore, this shadow price provides a good instrument to measure the pressure on cropland. Under all scenarios, the pressure on cropland decreases the most for the low-income region, since this is the region where most of the feed used in animal production is grown. For mid-income countries, this trend is also clearly visible, but the effect is much smaller, since this region is a large producer of ruminants, and therefore uses relatively little land for the production of feed. The effects of forcing animal diets shows that not taking these into account would underestimate the pressure exercised by the feed demand on land use: for all regions, the pressure on land is higher than that in the SHIFT and GREEN scenarios. For the EU, the effect is much less than that for the other high-income countries, since the higher preference for extensive agriculture led to a change in agricultural production structures that made the production of animals less dependent on commercial feed. Finally, including shifts in regimes in response to shifting income distributions dampens the results found under the DIET scenario somewhat, since, as was explained above, especially in the high-income region many people shift from the intermediate to the rich, satiated lifestyle.

GTAP

Whereas GEMAT focuses on specific and detailed developments, GTAP is used to model the general and international political framework. In the following, two types of simulations will be presented: the first includes general macroeconomic trends, which are independent of the introduction of

NPFs. The additional information concerning Gross Domestic Product (GDP) growth and trends in production factors was taken from Walmsley *et al.* (2000). The second describes future political developments relevant for the European Union (Brockmeier *et al.*, 2001). The scenarios were organized along a path which follows the expected time frame of future changes. The results of each scenario were used as the baseline data for the next scenario.
1. AG includes the Agenda 2000 reform package of the EU.
2. 2006 includes projections for the period 2002-2006.
3. ENLG includes an Eastern enlargement of the EU with the adoption of all border and domestic instruments including direct payments.
4. 2010 includes projections for the period 2007-2010.
5. CAP includes a possible next Common Agricultural Policy (CAP) reform with a reduction in direct payments by 30% and the abolition of milk and sugar quotas.
6. WTO includes a possible next WTO round with a reduction of all border instruments worldwide by 20%.
7. 2020 includes projections for the period 2011-2020.

The simulations constitute preliminary scenarios, which have to be defined in greater detail in the future. This becomes particularly relevant in regard to the distribution and diffusion of the new innovative product. However, they already provide a clearer vision about the relevance of different influences.

The results clearly demonstrate that the political developments considered in the scenarios only have a minor impact on the GDP. Here the changes result mainly from the macroeconomic projections. This is somewhat different for the production and consumption of (red) meat and NPFs, where both macroeconomic changes and politics may have a strong influence. Production and consumption of (red) meat will have their highest growth rates in the regions FNC and FIC (Food Neutral and Food Insecure Countries, respectively), where the rising GDP forms the main driving force, and the RCEE (Region of Central and Eastern European associate countries) where the accession to the EU provides additional incentives for producers. The Agenda 2000 leads to small changes in the distribution of (red) meat production and a minor total reduction in the EU (-0.9%). The enlargement results in additional regional shifts but increases the total output in the EU (+0.2%). The CAP reform and the WTO scenario include reductions in the political support of (red) meat production, which lead to a decrease in output. Hence, in general, the consumption of meat is only marginally affected by political changes. Here, again, macroeconomic developments dominate the results.

The macroeconomic projections lead to an increased output of NPFs in all regions. These results are based on the assumption that NPFs can be produced and consumed worldwide without any technological or political restrictions. As the invention of this product will take place in one region and will then follow a certain diffusion path, this may not be realistic. This restriction, however, was assumed to keep the simulations simple. Since no trade barriers or domestic measures were introduced for the NPFs, changes in the political regime only influence NPFs indirectly through cross-effects with other products, mainly land-using goods, particularly vegetables and processed vegetables. As already mentioned these scenarios are just a first step. Regarding the environmental background it seems quite realistic that the introduction of NPFs might be accompanied by a product-specific governmental policy scheme and/or policy measures designed to diminish the negative environmental effects of meat production. These can be included into the simulations to provide a more realistic scenario.

5.3.3 Conclusions

Different aspects of an economic problem may call for the use of different model frameworks. In this section, we concentrated on the effects of a transition from meat to NPFs, and showed, on the one hand, the importance of assumptions made with respect to agricultural production and consumption technologies and patterns (within the GEMAT model) and, on the other hand, the equal importance of representing institutional arrangements affecting agricultural production and trade (using GTAP). Of course, it is clear that on both accounts many improvements of the models are called for to arrive at a further improved representation of agricultural production, consumption, and trade.

It appeared that the GEMAT Full Format is very well suited for the inclusion of additional constraints on technology (such as animal diets) as well as for the representation of utility functions that diverge from the standard assumptions, extensions that we feel are essential for a correct representation of meat demand and agricultural production. GTAP can provide additional useful insights into the discussion about meat and meat substitutes, especially concerning macroeconomic developments, politics and, although not explicitly shown here, trade effects.

Due to the scarcity of information numerous assumptions had to be made. In addition, many extensions are possible that would greatly improve the description of agricultural production and consumption patterns within AGE models. Obviously, a very important omission is the lack of spatial specificity in the description of agricultural production, which requires integrating information on agro-ecological constraints on production within a spatially explicit economic model (as is done in, for example, Albersen *et al.* (2000)). Another important assumption of the modelling exercises is that the production of agricultural goods itself as well as the processing of these goods is carried out by fully competitive producers. This neglects the fact that, partly because of rising consumer concerns, agricultural production is increasingly being vertically integrated. This implies that market structure is changing and, in addition, that these vertically integrated conglomerates may try to control markets by engaging in strategic behaviour. In GEMAT this could be incorporated as described in Keyzer and Van Wesenbeeck (1999) and Van Wesenbeeck (2000), but it is not included here to keep the exposition clear and tractable. It is also quite realistic that in the beginning the newly invented NPFs will be produced by just a small number of companies, leading to monopolistic behaviour on the markets. This can also be included into the models (for the GTAP model, see e.g. Francois, 1998). The worldwide distribution and diffusion process of the newly invented NPFs can implement the GTAP model using the method developed by Van Tongeren and Van Meijl (1999).

In general, it is evident that a transition from meat to NPFs results in a decrease of environmental pressure on land under all scenarios (GTAP) and that the potential is even larger (GEMAT). However, an aspect of crucial importance to the feasibility of such a transition is the role of market prices. Interestingly, the two models employed disagree on future meat price development. From GTAP a price decrease may be inferred (Herok, 2003), but from GEMAT a price increase (Keyzer *et al.*, 2005; Van Wesenbeeck and Pavel, 2004). The underlying cause of this disagreement is deeply embedded in the structural differences of the two models, showing once more the added value of employing two complementary models.

5.4 INTERNATIONAL INSTITUTIONS [4]

5.4.1 Introduction

This section examines how the rules of international institutions may affect the marketing of NPFs. In particular, it examines the rules for market entry, promotion and support. The most important rule-setting international institutions in this respect are the European Union (EU) and the World Trade Organization (WTO). The question addressed in this section is whether

---

[4] Onno Kuik

international institutions provide mainly incentives or barriers for the introduction, marketing and promotion of novel foods and food ingredients.

5.4.2 Results

Market introduction and entry

Before a novel food can be placed on the EU market, it has to receive authorisation by the government. Since 1997, the procedure to get authorisation is specified in the Novel Food Regulation of the EU (Regulation (EC), No. 258/97). Novel foods and food ingredients that were already on the EU market before 1997 (such as pea protein) are not affected by this regulation, but new products must go to a lengthy and relatively expensive authorisation procedure. The Novel Food Regulation defines what is meant by a novel food or novel food ingredients:

- Which are produced from genetically modified organisms or which contain such organisms;
- Which present a primary molecular structure;
- Which consist of micro-organisms, fungi or algae;
- Consist of or are isolated from plants or isolated from animals;
- Whose nutritional value, metabolism or level of undesirable substances has been significantly changed by the production process.

In the authorisation procedure it is examined whether the novel food product presents a danger to the consumer, misleads him or her, or whether it is nutritionally disadvantageous to him or her compared with the product it replaces. The EU has issued guidelines for the scientific aspects of this examination (97/618/EC). The examination is initially carried out by a national competent body, but the European Commission (assisted by its Standing Committee on Foodstuffs) may overrule the national authorisation. The final authorisation decision specifies the scope of the authorisation and specifies the conditions of use, the designation of the food or food ingredient, its specification and the specific labelling requirements.

If the novel food is produced from or contains genetically modified organisms (GMO), it is subject to a special, additional procedure that emphasises the assessment of environmental risk. Public concern in Europe over GMOs and GMO food led to a de facto moratorium on new applications since 1998. The EU has recently tried to end this moratorium by approving new and strict legislation concerning the traceability and labelling of GMO food and feed, but this legislation has not yet been implemented (Regulation (EC), 2003).

The conditions of market entry are also of prime concern to the WTO and its members. The WTO administers international agreements on the conditions of international trade in goods and services. Its objective is to let international trade flow as smoothly, predictably and freely as possible. WTO agreements of relevance to NPFs include: the General Agreement on Tariffs and Trade (GATT) (establishing the basic principles of free trade), The Agreements on Technical Barriers to Trade (TBT) and Sanitary and Phytosanitary Measures (SPS), the Agreement on Subsidies and Countervailing Measures, the Agreement on Trade-Related Intellectual Property Rights (TRIPS), and the Agreement on Agriculture.

The WTO agreements specify the conditions of international market access. These conditions relate to import tariffs and quotas, but also to other rules and regulations that may unjustifiably discriminate between domestic and foreign suppliers, the so-called "non-tariff barriers". The members of the WTO are divided over EU policies on GMOs. The roots of the dispute between the EU and other WTO members (especially the US) are commonly ascribed to a different approach to the assessment of risk for human health and the environment. While the US would have a tendency to put the burden of proof on the prosecutor (i.e. the product is allowed except in the case of scientific proof of risk), the EU would have the tendency to put the burden of proof on the defendant (i.e. the product is not allowed except if it can be proved that there is no risk). Examples of this approach to risk assessment of the EU are the GMO and the hormones-in-beef import prohibitions. It has also been argued, however, that the main difference between the EU and the US in this respect is not so much a question of principle, but basically derives from differing socio-cultural tolerances for certain risks. While the EU would be less tolerant to risks related to GMOs and hormones, the US would be more sensitive to risks related to, for example, new drugs, blood donations, and mad cow disease. Anyway, the WTO rules favour the approach to risk assessment of putting the burden of proof on the prosecutor (allow "except"). The alternative EU approach to risk assessment (do not allow "except") might be a barrier to the introduction of NPFs.

Intellectual property rights

If an NPF is granted access to the (EU) market, the company that has developed the food may wish to establish exclusive rights over the sales of the products. These rights may be established by "intellectual property rights", if certain conditions are met. For example, for a patent to be granted the product or process should be "new, involve an inventive step and [be] capable of industrial application." The exclusive rights offer the developer of the new product monopoly rents for a specific length of time (often about twenty years). The size (and even the existence) of the monopoly rents will, of course, depend on the success of the product in the market.

The World International Property Organization (WIPO) administers most multilateral Treaties on Intellectual Property Rights (TRIPS). The TRIPS

Agreement of the WTO contains rules for intellectual property rights' regimes to which WTO members have committed themselves. If an NPF can be patented in one market (if it is sufficiently new, innovative and capable of industrial application), the TRIPS agreement offers its developers more certainty over the protection of their "invention" in overseas markets. The TRIPS agreement could therefore stimulate the invention and development of NPFs by offering international protection of the intellectual property rights of its inventors.

Subsidies and other government support

The national or EU government may wish to stimulate the sales of the product, for example because of its environmental superiority over the food it competes with. Forms of government support include direct subsidies and fiscal measures, such as a reduction of value-added taxes or other taxes on the product. These forms of government support are also subject to international rules, both at the level of the EU and at the level of the WTO.

EU principles on "state aid" are laid down in the EU Treaty and further developed in a number of regulations. State aid is, in fact, not allowed unless it falls under a specific exception. Whether a subsidy for the promotion of NPFs would qualify for such an exception is very uncertain.

WTO prohibits subsidies that are either designed to discriminate between domestic and foreign producers (prohibited subsidies), or can be shown to have an adverse effect on a foreign country's commercial interests (actionable subsidies). A study on the international trade aspects of the introduction of NPFs that was carried out for the PROFETAS programme (also see Section 5.3) showed that a Dutch consumer subsidy for NPFs might have the effect of an export subsidy (Herok, 2003). If this would be the case, foreign governments could file a complaint against such an "actionable" subsidy with the WTO.

A consumer charge on meat and meat products

A number of Dutch organisations have recently proposed a consumer charge on meat and meat products, for various reasons. The reasons included the generation of revenues for the destruction of cattle due to the BSE crisis, the generation of revenues to stimulate animal welfare, and the reduction of the price gap between organic and conventional meat. In practice, the consumer charge on meat and meat products could be implemented by an increase of the VAT rate from the present, reduced rate on foodstuffs (6%) to the general rate (19%), a specific consumer charge such as an excise tax, or a charge levied by the relevant commodity board ("*Productschap*").

A consumer charge on meat and meat products might also be considered as a way to stimulate the consumption of NPFs, either directly, by using the revenues of the charge to subsidise the production or consumption of NPFs, or indirectly, by its impact on the relative prices of meat and NPFs.

The Dutch government is not in favour of a charge on meat and meat products. Its arguments against such a charge are partly practical, partly economic and partly related to the international legal framework. With respect to the international legal framework, the following observations can be made:

- The EU rules on state aid do not allow a government to "earmark" the revenues of a consumer charge for a specific purpose (e.g. the destruction of BSE cattle). Such use of revenues would also be in conflict with the non-discrimination principle in international trade, because the supply of both domestic and foreign producers would be charged, but only the former would benefit from the revenue of the charge.
- "Meat" is not a well-specified product category. Meat comes in various forms and shapes and in various stages of processing into food products (such as pizzas). This poses problems for the VAT as well as for the excise tax alternative. Particularly for the excise tax alternative, the assessment of the meat content of an imported product may cause serious difficulties and may lead to unequal treatment of domestic and foreign suppliers. For the VAT alternative, the problem is that a product can only be in either the 6% or the 19% tariff rate (there is no middle way). If processed food would be included in the scheme, the increase from 6% to 19% VAT would be applicable to all food products with meat in them, however small the fraction of meat in the product.

It is important to note that international regulations do, in general, not prohibit taxes per se, but they do regulate the design of the tax. In general, the revenues of a consumer charge on meat and meat products should accrue to the general budget, and should not be used to finance certain measures in specific industries. In addition, the consumer charge should not discriminate between domestic and foreign suppliers.

An example may illustrate the kind of problems that arise because of the non-discrimination principle. Take the example that organic meat would be exempted from the consumer charge. The non-discrimination principle demands that organic meat from foreign suppliers should also be exempted from the charge. However, how can the national authority check whether all foreign meat that is supplied under the label "organic" is really organic? Specific measures should be taken to be able to certify the authenticity of the "organic" claim, and the non-discrimination principle would also require that the process of certification should be transparent and not be disproportionately difficult or expensive for foreign producers in comparison to their domestic competitors. All this could be difficult – and often it actually is difficult – to implement in practice.

5.4.3 Conclusions

Our examination suggests that even though international institutions provide both incentives and barriers for the introduction, marketing and promotion of NPFs, on balance the barriers exceed the incentives.

Market introduction and market support of NPFs are subject to certain rules. These rules are increasingly set at international levels: both at the EU and at the global level. Since 1997, the Novel Food Regulation of the EU sets the rules for the authorisation procedure. This procedure can be a barrier for the introduction of novel foods or food ingredients, especially if these foods or food ingredients are produced from GMOs. If this is the case, the authorisation procedure is very strict.

Once the novel food is granted market access, it is important for commercial developers of NPFs to have the exclusive rights over the sale of these foods and to be protected against imitation by competitors. "Intellectual property rights" regimes grant such rights and are increasingly based on multilateral cooperation and enforcement, through WIPO, the WTO and other organisations.

Direct government support to the marketing of NPFs is the subject of international, EU and WTO law. It is not clear whether consumer subsidies for NPFs would be allowed under the state aid provisions of the EU. If the consumer subsidy would have the effect of an export subsidy and would impose damage on the foreign production of NPFs, such a subsidy could be challenged before the WTO. An indirect way to stimulate the production and consumption of NPFs could perhaps be a tax on meat and meat products. A consumption charge would not be prohibited by international law per se, but international laws would certainly have an impact on the design of the charge and the use of its revenue.

The barriers erected by international institutions are mainly meant to protect the consumer and to resist protectionist practices in international trade. These barriers cannot be circumvented, whatever the potential qualities of the new product. For a successful introduction and marketing of NPFs, these barriers should be taken into account. Therefore, it is easier to start with foods and ingredients that have already been authorised than to start with foods and ingredients that still need to be authorised, in particular, if they are derived of, or contain, GMOs. For the promotion of NPFs, not too much should be expected of traditional government instruments such as taxes and subsidies. Subsidies have already lost their appeal in most EU countries for purely domestic reasons, and additionally they are heavily restricted by EU regulations concerning state aid and the single market. If NPFs are to become a success, it should primarily be through private, commercial means and actions. International institutions can protect and support commercial interests, for example, through the international protection of intellectual property rights.

5.5   CONCLUSIONS[5]

The main theme of this chapter is how a diet shift fits into the political and economic developments of the next decades. For the purpose of this treatise, a distinction was made between, on the one hand, the proximate decision makers in industry and government and, on the other hand, the members of society who participate in the market and the political processes. Decisions about food determine to a certain extent what kind of society people will get. However, the connections between their preferences and the performance of food systems are mediated by long and heterogeneous chains of interactions. The outcomes are partly a matter of political processes, partly a matter of international trade and world agricultural markets. In addition, the role of international institutions is increasing.

The chapter displayed several tools that the initiators or stakeholders of a new technology can use to monitor and influence the selection environment in which their products have to survive. This refers to scientific tools such as policy analysis and econometric modelling, as well as a skilful examination of institutional rules. Each of these tools provided relevant information about opportunities and barriers for NPFs. Under the present conditions of uncertainty, however, none could lead to definitive answers.

One of the general conclusions of this work refers to the role of pre-conceptions in thinking about the future. It is almost inevitable to use pre-conceptions when questions have to be answered about the desirability of policy options. But it appeared that preconceptions can lead to hidden assumptions that blur the arguments for or against an option. Section 5.2, for instance, noted a pre-conception that saw a diet shift primarily as an opportunity to develop products with a larger profit margin than meat. Another example is the implicit assumption mentioned in Section 5.3 that meat prices will continue to decrease in the future, just as happened in the past to most bulk supplies. Finally, Section 5.4 illustrated that not too much should be expected of traditional government instruments such as taxes and subsidies.

In contrast to what these pre-conceptions imply, it may be more sensible to develop plant protein ingredients that can serve as a low-priced alternative

[5] Joop de Boer

to meat protein ingredients. This was precisely the option proposed by the STD initiative (see Section 1.2). The TVP example mentioned in Section 5.2 demonstrated the consequences of neglecting consumers in product design. In addition, decision makers may show wisdom by expecting that meat prices will not decrease, but rather increase as a result of changes in the world market and agro-ecological constraints on production. Moreover, the proximate decision makers in industry and government may abstain from traditional policy instruments and facilitate the process of transition "management" instead (Section 1.1).

A critical reflection on pre-conceptions may also lead to the conclusion that the PROFETAS hypothesis about the desirability of a diet shift deserves some more stringent tests than the exploratory exercises that have been described in this chapter. Such a crucial hypothesis should not be confirmed too easily. A serious rebuttal, for example, could be that developing NPFs may contribute to increased production and consumption of proteins without a guarantee that the environmental pressure caused by meat production will be reduced. Another point is that the present analyses may have overstated the importance of agriculture for the political and economic processes in modern society. Alternatively, other linkages, such as with global energy issues, may have been overlooked. Therefore, Chapter 6 will explore arguments from a broader perspective.

At the end of the day, the very arguments that make a diet shift desirable (or not) will be different depending on the countries involved. What a transition from meat to NPFs may bring about is – in addition to more food sustainability (Chapter 2) – a more divergent food system with different approaches to food (Section 5.2) and a significant reduction of pressure on cropland (Section 5.3). Whether and to what degree these changes are desirable for a country cannot be answered without a further analysis of its prospects. How various arguments about the desirability of a diet shift can be combined is also country-specific. Nevertheless, several policy options, such as "develop new plant protein based alternatives" and "make it normal to eat less meat" are certainly not mutually exclusive. To put it simply, a combination of (a) eating less protein, (b) replacing a part of meat-based protein by plant-derived protein, and (c) replacing another part by extensively produced meat might prove optimal for any country's sustainability objectives.

REFERENCES

Albersen, P.J.J., Fischer, G., and Sun, L. (2000), *Estimation of agricultural production relations in the LUCC model for China*, IIASA Report IR-00-027, Laxenburg, Austria.

Brockmeier, M., Herok, C.A., and Salamon, P. (2001), "Technical and institutional changes in an enlarged EU: Welfare effects for old and new members with a focus on the agri-food chain", *Proceedings 4th Annual Conference on Global Economic Analysis, 27-29 June 2001*, Purdue University, USA.

Brown, N. (1999), "Xenotransplantation: Normalizing disgust", *Science as Culture*, Vol. 8, pp. 327-355.

Bruinsma, J. (2002), *World agriculture: Towards 2015/2030, An FAO perspective*, FAO, Rome.

Delgado, C., Rosegrant, M., Steinfeld, H., Ehui, S., and Courbois, C. (1999), *Livestock to 2020: The next food revolution*, Available at <www.ifpri.cgiar.org/index1.asp>, IFPRI/FAO/ILRI, Washington, DC.

FAPRI (2004), *Agricultural outlook 2004*, Available at <www.fapri.iastate.edu>, Food and Agricultural Policy Research Institute, Ames (Iowa), USA.

Francois, J. (1998), *Scale economies and imperfect competition in the GTAP model*, GTAP Technical Paper No. 19, West Lafayette, USA.

Glynn, S. (2002), "Constructing a selection environment: Competing expectations for CFC alternatives", *Research Policy*, Vol. 31, pp. 935-946.

Harvey, M., Quilley, S., and Beynon, H. (2002), *Exploring the tomato: Transformations of nature, society and economy*, Edward Elgar, Cheltenham, UK.

Herok, C.A. (2003), *The introduction of meat substitutes from vegetable proteins: Consequences for EU agriculture on a global level*, PROFETAS Project I3B Final Report, LEI, The Hague.

Keyzer, M.A., Merbis, M.D., Pavel, I.F.P.W., and Van Wesenbeeck, C.F.A. (2005), "Diet shift towards meat and the effects on cereal use: Can we feed the animals in 2030?", *Ecological Economics, (in press)*.

Keyzer, M.A., and Van Wesenbeeck, C.F.A. (1999), "Trade models of imperfect competition", in Ranis, G., and Raut, L.K. (Eds.), *Trade, growth, and development, essays in honour of Professor T.N. Srinivasan*, Elsevier Science, Amsterdam.

Lindblom, C.E. (2001), *The market system: What it is, how it works, and what to make of it*, Yale University Press, New Haven.

Mesaroviç, M.D., and Pestel, E. (1974), *Mankind at the turning point: The second report to the Club of Rome*, Dutton, Reader's Digest Press, New York.

Murray, F. (2004), "The role of academic inventors in entrepreneurial firms: Sharing the laboratory life", *Research Policy*, Vol. 33, pp. 643-659.

Nestle, M. (2002), *Food politics: How the food industry influences nutrition and health*, University of California Press, Berkeley, CA.

OECD (2004), *Agricultural outlook 2004-2013*, Available at <www.oecd.org>, Organisation for Economic Co-operation and Development, Paris.

Podolny, J.M., and Stuart, T.E. (1995), "A role-based ecology of technological change", *American Journal of Sociology*, Vol. 100, pp. 1224-1260.

Pyke, M. (1970), *Synthetic food*, John Murray, London.

Regulation (EC) (2003), *Concerning the traceability and labelling of genetically modified organisms and the traceability of food and feed products produced from genetically modified organisms and amending Directive 2001/18/EC*, EU, Brussels.

Van Tongeren, F., and Van Meijl, H. (1999), *Endogenous international technology spillovers and biased technical change in the GTAP model*, GTAP Technical Paper No. 15, West Lafayette, USA.

Van Wesenbeeck, C.F.A. (2000), *How to deal with imperfect competition? Introducing game theoretical concepts in a general equilibrium model of international trade*, Tinbergen Institute Research Series 231, Amsterdam.

Van Wesenbeeck, C.F.A., and Pavel, I.F.P.W. (2004), *Relieving the pressure: Alternative sources of protein for the year 2020,* PROFETAS Research Document, SOW-VU, Amsterdam.

Vijver, M. (2005), *Protein politics,* PhD thesis, Twente University, Enschede, The Netherlands.

Walmsley, T.L., Dimaranan, B.V., and McDougall, R.A. (2000), *A base case scenario for the dynamic GTAP model*, Available at <www.gtap.agecon.purdue.edu/resources>.

CHAPTER 6

EMERGING OPTIONS AND THEIR IMPLICATIONS

Harry Aiking, Joop de Boer, Johan M. Vereijken,
Anita Linnemann, Frank Willemsen,
Radhika Apaiah & Dick Stegeman

6.1 TOWARDS CROP-BASED SOLUTIONS[1]

The preceding chapters provide strong evidence for the notion that the use of meat protein in the world is extremely unbalanced, leading to a serious disturbance of the natural structures and biogeochemical processes on which life depends. To put it simply, in the developed countries, in particular, far too much meat protein is consumed to be globally sustainable. As a consequence, there are, on the one hand, excessive emissions in the meat producing countries themselves and, on the other hand, overtaxed natural resources (water, land) in countries that provide much of the meat and feed (developing as well as developed countries). Moreover, the situation is rapidly deteriorating as the world population continues to grow and increasing income in rapidly industrialising countries (China, Brazil) acts as a force driving up meat demand.

In view of the nature of this problem, crop-based solutions are called for. This should involve the development of products based on plant proteins that replace meat in a sustainable way. That is the key to the so-called protein transition. However, the PROFETAS projects have demonstrated that this solution will not just require the substitution of one type of protein by another. A satisfactory way of replacing meat proteins will require a whole package of options, which take into account how proteins are linked to other natural and societal issues. The protein transition can be realized only if it is based on a combination of linkages that will satisfy a whole set of constraints. The linkages include:
- crop choice (addressed in Section 6.2)
- envisioned use of by-products (6.3.1)
- consequences for other natural resources (water and energy) (6.3.2)

---
[1] Harry Aiking & Joop de Boer

- food-related issues conceived by key stakeholders in the market (6.4)
- issues put forward by governmental and non-governmental policymakers (6.5)
- contours of an evolving "global food economy" (6.6)

Although these linkages can be highlighted from three different angles (environmental sustainability, technological feasibility and societal desirability) this chapter aims to develop them in Sections 6.2-6.6 as a number of cross-cutting themes at a relatively general level in order to avoid being bogged down by too many details. Moreover, crop-based solutions will require a global vision that reaches far beyond the scope of PROFETAS.

As a first approximation, in PROFETAS the world was initially assumed to be rather homogeneous. From the results described in Chapter 2 it has become clear that, globally, the environmental benefits of a transition from meat to plant protein can conservatively be estimated to be on the order of a factor 3-10 for land and energy requirements as well as eutrophication, but even on the order of a factor 30-40 for water requirements and over 60 for acidifying pollution. To what extent and where, however, this might lead to decreased environmental pressure as a result of decreased pork production or, conversely, to increased environmental pressure as a result of locally increased plant protein production is less straightforward.

Among other things, the location of benefits or detriments strongly depends on the actual crop choice (Section 6.2). This crop choice also depends on future developments in the areas of texture formation, protein-flavour interactions and plant breeding (Sections 3.2-3.5). Furthermore, crop choice is complicated by the emerging fact that the protein transition cannot be realistically uncoupled (Section 6.3) from both the biomass transition (towards sustainable energy production) and the water transition (towards sustainable use of freshwater).

From a societal point of view the question should be raised how the future of NPFs will be decided when decision makers in industry and government are considering the issues on their agenda. Commitment by the actors who are important for the prospects of NPFs is the theme of Section 6.4. The analysis of actor commitment has been elaborated with a section on actual feedback from Dutch actors (Section 6.5). The final part of this chapter puts the main findings into the broader perspective of an evolving global food economy (Section 6.6).

6.2 CROP OPTIONS[2]

6.2.1 Introduction

The major crops from which proteins are derived for application as an ingredient in man-made food products are soy and wheat, soy being the most important one. Next to protein preparations, the processing of these crops yields two other commercially valuable bulk products, i.e. starch from wheat and oil from soy. In fact, wheat and soy processing have been invented to obtain these latter components. However, especially for soy the isolation of protein preparations has become more and more important from an economic point of view.

The protein preparation derived from wheat (termed wheat gluten) is mainly used in the baking industry, among others to enhance the baking quality of wheat doughs. From soy, a number of protein preparations are produced: defatted meals, concentrates and isolates, with a protein content of about 50, 70 and over 90%, respectively. There are numerous applications for soy protein preparations as ingredients in human food: they are used, for example, in soups, desserts, dressings, bakery products and, last but not least, processed meat products.

On a much smaller scale other crops are used to produce protein preparations to be used as ingredients in human food. Examples are pea, lupine and rice. In the feed industry many other plant protein preparations are used. These protein preparations are mainly derived as by-products from oil (rapeseed, sunflower) or starch processing (corn, potato). However, soy protein preparations dominate in the feed industry, too.

The major source of protein for the production of meat substitutes is soy (Section 3.1.2). However, due to climatic causes, soy is only grown on a very small scale in Europe. Because the PROFETAS programme aims at developing NPFs from proteins derived from crops that are or can be cultivated in Western Europe, a different crop should be selected. To this end a desk study was performed before starting the actual PROFETAS programme. The results of this study are described in the next section. Subsequently, the choice of the crop will be evaluated in the light of the

---

[2] Johan M. Vereijken & Anita Linnemann

results obtained within the PROFETAS programme. Finally, other options will be discussed.

6.2.2 Desk study

Before the PROFETAS programme actually started, a desk study was performed in which eight crops were compared with respect to their suitability as a protein source for the development of NPFs. These eight crops were selected because they can be cultivated in temperate climates such as in Western Europe and have an assumed potential to be the protein source of NPFs. This potential was based on the economic perspective for commercial protein production, the functionality of the proteins and the relatively low impact on the environment by the cultivation of these crops, as derived from discussions with crop specialists, technologists and environmental experts.

The crops studied are two legumes (lupine, *Lupinus* spp. and pea, *Pisum sativum*), two non-legume grain crops (quinoa, *Chenopodium quinoa* Willd. and triticale, x *Triticosecale*), two leafy crops (lucerne, *Medicago sativa* and grass, *Lolium* and *Festuca* spp.) and two other crops, rapeseed (*Brassica napus*) and potato (*Solanum tuberosum*) that are being processed to yield starch (potato) and oil (rapeseed). From all these crops the seeds are the source for protein production except for the leafy crops and potato; from these crops leaves and tubers are the source, respectively.

The desk study was split into two parts, which have been published separately (Dijkstra *et al.*, 2003; Linnemann and Dijkstra, 2002). The aim of the first part was to analyse and evaluate the primary links of the production chain. In the second part, the focus was on technological issues of the production chain.

6.2.3 Primary links of the production chain

In the first part, the perspectives of the eight crops as possible sources of protein for NPFs were analysed with respect to the primary links of the production chain. The suitability of the crops is determined by classifying them with regard to the aspects most relevant to these links. The aspects taken into consideration were familiarity of farmers with the cultivation of the crop, perspectives for rapid crop improvement, protein production (kg/ha), protein quality (absence of unwanted substances) and familiarity with usage for human food in Western Europe. The classes used were + = moderately good, ++ = fairly good and +++ = good.

Familiarity of farmers with the cultivation of the crops

Some of the crops are already being cultivated in Western Europe on a large scale and hence have the advantage that the farmers have experience with them. These crops are pea, triticale, lucerne, grasses, rapeseed and potato. The suitability of these crops was therefore classified as good (+++). Lupine has been cultivated more extensively in the past than nowadays. Quinoa, however, has no history of cultivation in Europe. The suitability of the latter two crops was thus classified as fairly good (++) and moderately good (+), respectively.

Perspectives for crop improvement

The crops differ significantly in the basis that is available for crop improvement. This basis includes genetic materials, scientific knowledge and infrastructure. Not surprisingly, the ranking of the crops with respect to this aspect is similar to that of the familiarity of farmers with the cultivation of the crops.

Protein production in kg/ha

Large differences exist in the amount of protein that the crops potentially yield per hectare (see Table 6-1). A subdivision into three groups was made as follows: (a) good (+++): lucerne and grasses (> 2000 kg/ha), (b) fairly good (++): pea and lupine (1000-2000 kg/ha), (c) moderately good (+): triticale, quinoa, rapeseed and potato (< 1000 kg/ha).

Table 6-1. Crop and protein yield.

Crop	Yield (ton/ha)	Relevant raw material	Protein content (%)	Potential protein yield (kg/ha)
Pea	3-5	Seed	25	1250
Lupine	3-5	Seed	33-40	2000
Triticale	6	Seed	11-14	800
Quinoa	3-4	Seed	14-16	650
Lucerne	8-14	Leaves	20	2500
Grasses	10-16	Leaves	20	2500
Potato	45	Tuber	1	450
Rapeseed	3.5	Seed	19-22	700

Protein quality

None of the crops in this study yields relevant raw materials (seeds, tuber or leaves) that are completely free of antinutritional factors (defined as factors that have a negative effect on nutritional quality such as phenolic compounds, enzyme inhibitors, (glyco)alkaloids) and/or poisonous substances. However, for pea and lupine, food grade flours and protein-rich products are commercially available, indicating that the quality of the protein is good enough for further processing into human foods. Pea and lupine are therefore classified as good (+++), where the other crops are classified as fairly good (++).

Familiarity with usage for human food in Western Europe

Some of the crops have a long history of usage in the diet in Western Europe, but others are completely unknown. The products prepared from unfamiliar crops will require an additional effort in time and money from marketing divisions, before their usage is well accepted among consumers. This implies that the acceptability of protein from pea and triticale is expected to be good (+++), and that of lupine fairly good (++). Introducing protein from quinoa, lucerne, grasses, rapeseed and potato in the diet of consumers will need more explanation and advertising (+).

Judgement

The suitability of the eight crops for the production of protein in Western Europe was judged by weighing up the pros and cons of the elements of the primary production chains that were studied. The result is that pea, lucerne and grasses are the most promising (+++). Fair prospects (++) are foreseen for lupine, triticale, rapeseed and potato. The possibilities for quinoa (+) lag significantly behind those of all the other crops.

6.2.4 Technological aspects of the production chain

In the second part of the desk study the emphasis was on the technological aspects of the production chain. This includes the processing technology, and functional and nutritional qualities of the derived proteinaceous products.

Processing methods

For all eight crops methods are available to process the eight raw materials into protein-rich products such as protein concentrates and isolates. However, there is much less experience with the processing of quinoa than with the processing of the other crops. The technology to isolate potato protein, which is now used by industry, results in a preparation consisting of denatured and strongly aggregated protein, which can only be used for nutritional purposes in feed. Pea and lupine are used industrially as sources for the production of protein concentrates and/or isolates intended to be used in human food.

Functional properties

The term functional property has been defined as "any physicochemical property which affects the processing and behaviour of protein in food systems, as judged by the quality attributes of the final product" (Kinsella, 1976). Examples of such functional properties are emulsifying, foaming, gelling and water-holding properties. The set of functional properties which a given protein preparation exhibits, depends on the nature of the proteins present (i.e. it depends on the plant species and, usually to a lesser extent, within the species, on the variety), but also very strongly on the technology by which the protein is processed. Furthermore, functional properties are affected by factors such as protein concentration, pH, ionic strength, temperature and the presence of other components such as carbohydrates. In addition, some properties are variety-dependent. Finally, different laboratories use different tests and/or conditions to assess functional properties. In conclusion, it is virtually impossible to compare functional properties of protein products in a scientifically satisfactory, reliable manner.

Nutritional properties

For similar reasons as for functional properties, literature data on the nutritional properties of protein preparations are hard to compare. Furthermore, to assess these data the relevant question is whether one wants to aim at the highest possible nutritional value, or, in contrast, whether one is satisfied with a proteinaceous product that has no traceable amounts of antinutritional and/or poisonous components. To be a source for the production of NPFs, it is essential that a suitable protein source can deliver sufficient protein without these latter components to a yet to be defined end product. This can be achieved with all eight sources.

Judgement

It was concluded that the technological possibilities of the eight crops cannot be used to discriminate between their suitability as a starting material for NPFs. Pea and lupine have a slight advantage over the other crops, because their concentrates and isolates are already commercially available.

6.2.5 Overall judgement

Based on the desk study, pea, lucerne and grasses have the highest potential as protein sources for the production of NPFs. Among these three, pea has the slight advantage that industrially protein preparations are already being produced. Because one model crop should be selected, the pea was chosen. Next to the elements considered in the desk study other elements were taken into account, the most important one being the resemblance, both with respect to biochemical and functional properties, between the major pea and soy proteins. Compared to other plant proteins, a lot of literature is available about soy proteins with respect to production, functionality and applications. This includes literature on aspects that are of the utmost importance for the production of NPFs such as thermal behaviour and texturisation.

It should be kept in mind that despite the desk study, the choice for pea proteins as model proteins to develop NPFs is at least partly arbitrary. Not all elements of the production chain have been taken into account, for instance, the environmental aspects (see Section 2.4). Furthermore, no selection could be made based on the information available with respect to the technological aspects of the production chain.

6.2.6 PROFETAS programme results

With respect to the choice of the crop to be the starting material for the development of NPFs, the PROFETAS programme did not yield decisive additional results. It should be emphasised that none of the projects was intended to contribute to this choice, but they were intended to deliver tools to develop NPFs and they succeeded well in this respect. Nevertheless, it is worthwhile to discuss their results in the framework of crop choice for NPFs.

Two projects are directly linked to the primary production, i.e. the projects on crop growth (see Section 3.4) and genetic modification (Section 3.5). Both projects yielded new technological instruments: a new method for genetic modification of peas and an innovative crop growth model. The latter model is generic and can be used for any arable grain/seed crop. The newly developed technology for genetic modification of pea may also offer new perspectives for the genetic modification of other legumes such as soy. However, it is not possible to compare this technique with techniques for genetic modification of other crops in terms of crop choice (e.g. required time, effectiveness and efficiency). To this end more experience with the newly developed technique is required.

Two other projects are directly linked to functional properties of pea proteins: NPF texture formation (Section 3.3) and NPF flavour retention (Section 3.2). The main objective of the latter project was to get insight into protein-flavour interaction using pea proteins as model proteins. It succeeded well, but with respect to crop choice, it did not yield additional discriminating information, for the results are largely generic and – except concerning the relatively pea-specific saponins, maybe – the insights can be used for other protein sources.

The project on texture formation yielded some information concerning crop choice. The results appear not to be directly in favour of pea protein.

Peas contain a protein fraction (a part of the vicilin fraction) that hampers the gelation at near neutral pH. However, a similar protein fraction is present in other seeds (e.g. soy), too. More discriminating is the observation that another pea protein fraction, legumin, is inherently less able to form a well-structured gel network than its corresponding counterpart in soy, i.e. glycinin. Furthermore, in contrast to soy glycinin gel networks, those of pea legumin were found to be susceptible to rearrangements during re-heating that caused gel strengthening. This makes the behaviour of pea legumin gels less predictable when re-heating is required (as, for instance, for preparing NPFs at home). So pea legumin seems to be less well suited for NPF production than soy glycinin. However, legumin is only one of the protein fractions present in peas and combinations of the protein fractions (as in commercially produced pea protein isolates and concentrates) might behave in a way that is more comparable to corresponding soy protein combinations. Furthermore, differences in gelling behaviour, despite its importance in texturing, might not result in differences after texturing the proteins by techniques other than gelling, such as extrusion.

Another argument pleading against the use of peas derives from the fact that pea processing will also yield starch. The amounts of pea required to produce enough protein to replace 40% of present meat consumption will result in an amount of starch that nearly equals present global starch production. In contrast, the use of the protein fraction from oil seeds such as soy will result in an amount of oil that equals about 10% of the present oil production. However, it is certainly not excluded that a major part of the pea starch will be used for the production of NPFs. This type of issue will be dealt with in more detail in Section 6.3.

6.2.7 Other options

At the start of the PROFETAS programme, peas were selected as the model crop to derive proteins for NPF development. Most of the arguments are still valid. Rapeseed, however, especially the "double low" varieties called canola (low in two antinutritional factors of rapeseed) are now becoming the starting materials for the commercial production of concentrates and isolates. This leads to a higher ranking of this crop.

Based on the results of the PROFETAS programme, especially those on texture formation (Section 3.3), one could argue that crops having the legumin type of protein and lacking the vicilin type of protein might have an advantage over peas, since one of the vicilins is hampering gelation. Examples of such crops are sunflower (*Helianthus annuus*) and rapeseed/canola, both familiar oil seed crops. Processing these crops yields oil instead of starch, which might also be advantageous. However, the proteins from these oil seeds also comprise, in relatively high amounts, a type of storage proteins not present in pea or soy, so-called 2S albumins. The presence of this type of protein will most likely affect the behaviour of the legumin type of proteins and, hence, have an effect on properties relevant for NPF production, taste and texturing. This would evidently require further investigations.

For crop choice the other components that are or can be obtained during processing of the relevant parts of the plant (seed, tubers or leaves) should also be taken into consideration. As already mentioned, crops that also yield oil may be favoured over crops that also yield starch. This would rule out cereals such as wheat and corn and tubers such as potato. During processing also fibre fractions will be obtained. At present, these fractions have little applicability, just like the parts of the plant that are not being used, such as straw. However, things may change. For instance, in the framework of sustainable energy production a lot of research has been performed, aimed at evaluating the use of fibres, starch and oil for the production of sustainable energy. Especially, with respect to the use of oil and starch for production of biofuels (bio-ethanol and biodiesel) a lot of information is already available.

For crop choice, environmental aspects should also been taken into consideration. Next to issues such as amounts of pesticides and fertilisers used for cultivation, water usage for both cultivating and processing crops is likely to become an important parameter for crop selection (see Section 6.3). In addition, agronomical differences between crops should be considered. Such agronomical differences include cropping systems (crop rotation, mixed cultivation of two crops) and sensitivity towards biotic stress (resistance to pests and diseases) and sensitivity towards abiotic stress (drought resistance, tolerance to high and low temperatures). Such cultivation issues will likely affect crop choice because they contribute to consistency of the yield (and hence income of the farmer) and to consistency of the crop supply (hence, attractiveness for food producers and processors, respectively).

In this respect, when talking about crop selection, it may be considered that molecular biology (biotechnology) might contribute to improving crops as a source for NPF production. This may not just concern the proteins themselves (their composition, ratio of protein types, amounts; see Sections 3.5 and 3.8), but also the plant as a whole. For instance, genetic modification might conceivably contribute to an improved straw stiffness for plants such as peas, resulting in a better resistance to lodging. Although at present the technological feasibility and social desirability of such developments is unclear, such technological achievements might result in higher yields and better suitability of crops to be a source of NPFs.

It should be kept in mind that NPFs need not be produced from proteins derived from a single crop. It may well be that the required combination of functional properties and the performance with respect to each of these functional properties cannot be met by proteins derived from one single crop. As an illustration, some of the present meat substitutes are based on two crops (wheat gluten and soy proteins, in addition to significant amounts of egg protein).

Last but not least, it is to be expected that, primarily due to soil and climate conditions, different crops will be used in different regions of the world. For instance, in warm regions of Asia soy is a good candidate, delivering both oil and proteins. Furthermore, soy based NPFs may suit Asian consumer preferences better than in Europe, because soy protein products such as tofu and tempeh have been part and parcel to the Asian diet for centuries. In Africa, groundnuts might be an option, since it is a familiar crop and delivers both proteins and oils. A legume that is also quite familiar in both Africa and Asia (especially India) is chickpea. However, as with peas the main non-protein constituent is starch. As already argued, in more temperate regions, such as Europe, Canada and Australia oilseeds such as rapeseed and sunflower might be interesting options.

6.2.8 Conclusions

At the start of the PROFETAS programme, peas were chosen as the model crop for the development of a technological toolbox to produce NPFs. Most of the arguments for this choice are still valid. However, the results of the PROFETAS programme suggest there are other options as well. Under the present conditions – which could easily change as a result of changing world market prices or newly emerging technologies – in Europe oilseed crops seem to have an edge, particularly, certain oilseeds lacking a vicilin type of protein. However, it should be emphasised that selecting potential crops for NPF production was not a PROFETAS priority. Such would require dedicated research regarding differences in cultivation aspects (both with respect to agronomical and environmental issues), as well as developments in breeding. Furthermore, outside Europe other crops are likely to be used as the starting material for NPFs production.

Looking into the future, other developments, for instance those aimed for in programmes concerning sustainable energy production and water usage, may well affect crop choice. Therefore, links should be established with relevant research programmes in those and other transition areas.

6.3 COMBINED CHAINS[3]

6.3.1 Introduction

The chains that have been chosen as models for the protein transition studied in PROFETAS are the pork chain and the pea-NPFs chain. Neither of these chains exists as an independent entity, for they require inputs from other chains and produce inputs for other chains. When looking towards future options it is important to note that lessons can be learned from studying the pork and pea chain, but these chains serve only as models and other chains may be more suitable for a protein transition.

Although the focus of the programme is primarily on Europe, it is inevitable that when dealing with transitions of the magnitude envisaged, effects on a global scale are taken into account. This is especially true when looking at uses of – from the perspective of the PROFETAS programme – by-products. In small volumes these products may be sold on the basis of a specific characteristic of that by-product, but on a larger scale it is easily conceivable that any niche markets will be flooded. A considerable amount of by-product will have to be sold in bulk, based on "bulk" characteristics. Section 6.3.2 will be devoted to the linking between the (meat and pea) protein chains, on the one hand, and agricultural input chains and by-product output chains, on the other hand. Section 6.3.3 sketches the potential consequences of altogether replacing feed crops and the opportunities this might entail for the global environment.

6.3.2 Proteins and by-products

As the PROFETAS programme studies a transition from meat protein towards plant protein there are specific trends that need to be distinguished:

(A1) A decrease in meat production means a decrease in inputs into the meat chain. This should not pose a major problem when crop products such as seeds are concerned, but it may incur environmental and monetary costs where wastes from agriculture and from the food industry are currently fed to pigs.

---

[3] Frank Willemsen, Radhika Apaiah, Dick Stegeman & Harry Aiking

(A2) A decrease in meat production means a decrease in outputs from the meat chain. This would lead to a decrease in such varied by-products as manure, fat, sausage casings, felting, leather, gelatine, etc. In the West-European situation a decrease in manure would probably be classified as a positive environmental effect.

(B1) An increase in protein crops grown would mean an increase in inputs into the protein crop chain. Depending on the cropping method and the crop involved this will result in an increase in the use of water, land, fertilisers, energy and pesticides.

(B2) An increase in protein crops grown would mean an increase in non-protein output from the crop chain, such as oil, starch or cellulose.

In the following paragraphs the interactions between different chains will be discussed, as well as possible substitution between chains. A decrease in available animal fat for example, may be offset or even more than offset by an increase in available vegetable oil. Likewise, the increased land use for protein crops may or may not be compensated by a drop in land use for feed crops, etc.

(A1) Effects of a decrease in meat chain inputs on other chains

In the PROFETAS programme the pig chain was chosen as the meat protein model chain. Although the meat chain uses other inputs such as water, energy and antibiotics, we will focus on the feed use of the pig chain. An important characteristic of pigs is that, unlike cattle and sheep, for example, they are unable to digest cellulose and, therefore, require approximately the same kind of nutrients as humans.

In Western Europe pigs are fed a complex mixture of seeds from purpose-grown crops and wastes from various stages of other chains that lead from crop to end product for human consumption. Presently, both parts of this mixture are about equal in size, but it is important to take into account that soy, for example, used to be grown primarily for its oil, whereas nowadays its protein-rich feed ingredients have gained importance. Another noteworthy trend is that, as a result of the increasing level of quality that consumers expect, more food that is deemed fit for human consumption will be rejected and that this food will often end up in pig feed.

If demand for inputs into the pig chain decreases, the producers of feed crops will have to adapt. For purpose-grown crops, this may lead to any or all of the following effects, depending on price and the choices available to the primary producer (farmer) of the crop:
- The producer continues to supply the pig feed market albeit at a lower price.
- The producer supplies a different market with the same crop.
- The producer supplies a different market with a different crop.

- The producer stops growing crops.

All except the first option are potentially interesting as far as a protein transition is concerned. The "different market" for crops that are currently grown for pig feed could be the food market and such a crop may be suitable as a source of NPFs. It also may be the industrial market (technical products ranging from adhesives to antibiotics, binders, cosmetics, inks, paints, etc.) in which the crop may be used to replace by-products from the pig chain.

The "different crops" can be crops for the (protein) food market and/or for biomass. Land being taken out of production is under the current pressures not very likely, but as the land becomes available, it could decrease the pressure on land use elsewhere.

For the "wastes" the situation is slightly different. As the price of the waste streams goes down, there are several options available:
- Industries may improve the efficiency of their processes, resulting in less waste.
- Industries may choose to have their waste processed differently.
- Industries may cease production of the main product from which the waste stream originates.

In the latter case it is difficult to see links with the protein transition. The effects however, would need careful study in terms of sustainability and social desirability. An improvement of the efficiency of a production process, for example, will usually result in environmental benefit, but processing waste differently, e.g. land filling or incineration, may be detrimental to the environment. Processing waste in a sustainable way, therefore, is an important discipline for study, which is also acknowledged by policymakers. The European Commission financially supports research on further processing by-products into higher added value products. Ceasing the production of certain products may or may not be more sustainable and social desirability must definitely be a question here. A final remark on the use of waste in animal feed is that using waste streams in this way has recently been reduced, because of the risk of contaminating the human food chain through unsuitable wastes. This development is putting great pressure on the food industry, and particularly the meat industry, because due to the BSE crisis the use of animal waste streams (including swill) is bound by many restrictions and may not be used for feed products for the same species anymore.

(A2) Effects of a decrease in the meat chain outputs on other chains

The meat chain produces a lot of associated outputs in different stages of the chain. First of all, there are outputs during the growth of the animals, mainly manure. In the Western European situation, where fertiliser of artificial and animal origin are both abundant, a decrease of these outputs is unlikely to have an effect on other chains, or it would have to be a marginal increase in artificial fertiliser use.

Second, during slaughter several by-products are produced that are subsequently used to produce leather, gelatine, glues, etc. It would be very difficult and time-consuming to have to consider all these products separately. Although they represent a considerable volume, at this stage it is difficult to say how much of an effect on other chains would be noticeable. However, it is assumed that in almost all applications good, but perhaps more expensive, alternatives for the meat by-products are available.

Third, a category of wastes emerges either during slaughter or in further processing afterwards. These consist of cut-offs, unpopular cuts of meat and fat. Currently a common destination for many of these is the pet food industry. Here, also, it is unclear what kind of effect a decrease in this sector will have on other chains. In terms of the energy value and chemical structure, animal fat is comparable to most vegetable oils (both are triglycerides), so that a decrease might be countered by an increase in, if necessary chemically hardened, vegetable oil production.

(B1) Effects of an increase in inputs to the protein crop chain on other chains

Linked with a considerable increase in growth of protein crops are increased land and water use and (depending on cropping system and crop) increased fertiliser use, increased use of agricultural chemicals and fuel. Such increases will primarily cause a greater demand for land, water, etc. and thus affect other chains. Considering the rationale of the PROFETAS programme, the increases should be offset by corresponding decreases in feed crop production. Because of the inefficient conversion of plant protein into meat protein (on average 6 : 1) the total amount of protein crops produced will, no doubt, decrease by the transition. In many cases feed crops are comparable to protein crops for human consumption so that it would not be difficult to assess the environmental impact. As the plant protein chain in the PROFETAS programme is to a large extent virtual, it should be possible to choose the protein crop so that it is a very close match with feed crops. The consequences of choosing different protein crops for NPF production will have to be studied more intensively.

(B2) Effects of increased outputs from the protein crop chain on other chains

Although the plant protein foods that are the object of study in PROFETAS are rich in protein, processing steps will inevitably yield one or more fractions low in protein, but high in non-protein substances (depending on the crop, primarily either starch or oil), in addition to the desired protein fraction (see Section 3.7). If a large part of the world's meat protein consumption were to be replaced by NPFs, the sheer volume of the non-protein fractions that would arise is immense. This rules out any niche markets for the non-protein fractions and so they would have to compete with other bulk agricultural products on a global scale.

To understand the effect of the extra non-protein fractions we may compare the present volumes of these fractions with potential volumes if 40% of the meat consumption (equivalent to about 18 million tons of protein) is replaced by plant protein consumption. Taking peas as an example of a possible protein rich raw material containing much starch, the production of 18 million tons of protein will come with the production of about 36 million tons of starch. This is on the same order of magnitude as the present global starch production (i.e. 48.5 million tons in 2000). Thus, using peas (with the present composition) as the protein source will affect the present market situation severely.

If we look into more detail at the oil fraction, soy may be taken as an example. The world consumption of vegetable oil is over 90 million tons now (of which the larger part is used in food production). If soy is used to supply 18 million tons of protein, about 9 million tons of soy oil is liberated. This is about 10% of the present consumption and, therefore, no severe long-term effects were to be expected if it were extra production. Interestingly, a protein transition of this sort would result in a shortage of soy oil, rather than in a surplus, as will be seen in the next section.

For the non-protein fractions, applications are likely to be food, feed, industrial raw materials and energy. Use as food (and due to the volume we may say staple food) is for some crops already a reality, for example, if we look at soy, where the oil fraction is used as cooking oil and as an ingredient for a wide variety of food products.

Considering the rationale of the PROFETAS programme (avoiding inefficient animal conversion for reasons of sustainability), use as feed is only an option for fractions high in non-starch polysaccharides, especially if no other use for this fraction is developed. Use as feedstock for the chemical processing industry is currently under investigation for various stocks. In fact, a trend towards using all components of a crop – coined the "biorefinery" concept by analogy to mineral oil fractionation – has recently emerged. Use for retrieving energy, although currently not a very economical option, is interesting because of the possibility of producing CO_2-neutral fuel. Both oil and starch are used to produce the automotive fuels biodiesel (resembling diesel) and ethanol (resembling petrol), although the production of such biofuels currently needs subsidies to be viable in the market. This option will be dealt with in the following section.

6.3.3 Opportunities for the global environment

In the preceding section the changes in the meat protein and plant protein chains were assessed separately. But if one takes a larger perspective and looks at the position of protein crops in world agriculture, food and feed crops are closely related. As the PROFETAS programme assumes the replacement of meat with plant protein based products, a useful way to look at this relation is to look at the total current area of feed crops and assume that by discontinuing the production of animal feed altogether this land will become available for a protein crop that can then be used as the basis for an NPF. The land that is left over can then be used for any other use (crops, nature). Please note that this is an extreme scenario aimed at examining maximum values for changes in land use. Therefore, little consideration is given to economic and institutional constraints that no doubt would have a role to play if such changes were actively pursued.

The amount of land currently used as cropland is about 1.5 Gha (=15 million sq. km). Of this area around 400 Mha is used for growing feed crops (FAO, 2005; OECD, 2004a).

Table 6-2. World land area and its uses (FAO, 2005). (* Figures for "Forest and woodland" and for "All other land" are from 1994, the last year in which FAO included these categories in its agricultural land use statistics. All other figures refer to 2002, the most recent year included by FAO.)

Land use type	Current Area (10^6 ha)	
Arable & permanent cropland	1,500	
of which for feed		400
Permanent pastures	3,500	
Forest and woodlands*	4,200	
of which managed forest		500
All other land*	3,900	
Total	13,100	

From Table 6-2 it is not immediately clear that there is any shortage of arable land. But any expansion of the arable land area would have to come from either the "Permanent pastures", "Forest and woodlands" or "All other land" categories. The "All other land" includes deserts, mountain ranges but also built up areas, and cannot be used for cropland. Converting land that is in the "Forest and woodlands" category would probably not be considered desirable, although a likely development for some areas, as it will encroach upon areas that up till now had been relatively unspoilt, and which are reserves of biodiversity. What is reported as "Permanent pastures" may look promising, but these lands are often used as pasture for the very reason that they are unsuitable as cropland. This is mentioned explicitly, for example, in

"World Agriculture: towards 2015/2030" (Bruinsma, 2002). Therefore, we will assume in our calculations that the current land use is a fair estimate for what is the most suitable land use.

The main crops grown for feed are grains and oil crops of which the grains are most often used entirely as feed whereas oilseeds are crushed and only the resulting "cake" is fed to animals, the oil being mostly used for human consumption. What options there would be if the feed area were used for food is displayed in Table 6-3 below.

Table 6-3. Comparison of alternative uses of the feed area. Sources for "Present": areas and amounts from De Haan et al. (1997) with oilseed area attributed to feed crops estimated on a weight basis, and protein contents from FAO (2004) and Berk (1992); for "Pea": conservative estimates for yield and protein content from Linnemann and Dijkstra (2002), and fractions resulting from air classification from Tyler et al. (1981); for "Soy": a medium high yield value from FAO (2005), cake and oil amounts from De Haan et al. (1997), and cake protein content from Berk (1992).

	area (10^6 ha)		amount (10^6 tons)	products	use
Present	100	oilseeds	70	oil	human
			140	cake	livestock
	300	grains	800	grains	livestock
Pea	52	pea	53	protein fraction	human (NPFs)
			103	starch fraction	any
	348	other uses			any
Soy	25	soy	15	oil	human
			58	cake	human (NPFs)
	375	other uses			any

Table 6-3 clearly shows that the feed route supplies protein for human consumption rather inefficiently. Thus, 800 Mt of grains are presently produced which, assuming a protein content of some 10%, provide 80 million tons of feed protein, plus 140 Mt of oilseed "cake" providing another 64 Mt of protein. With an estimated production of 144 Mt of feed protein, at the very most 29 Mt of meat protein can be produced, assuming a conversion efficiency of 20% based on Smil using USDA long-term statistics for poultry (Smil, 2002). In reality, however, not all the protein will be fed to poultry and therefore the efficiency will be lower still.

Alternatively, the production of 29 Mt of pea protein for human consumption would require only 52 Mha of land (based on separation into two main fractions by means of air classification). This would create a starchy fraction (not pure starch) of 103 Mt. Such a quantity of starch would be very difficult to market as a bulk product as it would have to compete with starches that are produced more cheaply from higher yielding crops. However, 348 Mha would be freed to be used for food production, biomass for energy, nature conservation or other uses, depending on societal developments.

Producing 29 Mt of protein from soy would require about 25 Mha of cropland. In this case, the major side-product is oil, which can be used for human consumption without any difficulty. Use as a stock for industrial products is currently a niche market, but use as starting material for biofuel is increasing, due to a variety of factors. Just as in the "pea" scenario, the remaining 375 Mha could be devoted to any other use.

It should be noted that under the current production a considerable amount of oil for human consumption is produced, but this would be reduced (from 70 Mt to 15 Mt) in the case of soy for direct human consumption and no oil is produced at all if peas are grown for protein. This shortage might pose a considerable problem, initially, but not one that cannot be redressed. More compelling may be the question whether there will be a demand for the huge starch fraction that is left when peas are grown to replace meat protein. So pea production for protein replacement frees up less land than the soy option, provides an even larger gap in the supply of vegetable oil and yields a potentially problematic amount of starch. Although the scenario presented is used to examine extreme changes in land use world-wide, it is interesting to further complete the picture by paying attention to some marked benefits and some consequences for the meat sector and consumers.

Using the freed-up area for biomass could provide the first enormous benefit. Assuming a biomass yield of 15 tons/ha of dry mass, a value that we believe is realistic (Van den Broek, 2000), the remaining 375 Mha would yield $5.6*10^9$ tons of dry biomass. Using a higher heating value of 19 GJ/ton (Hoogwijk *et al.*, 2003), this amount of biomass could provide 100 EJ of energy, over 25% of the current energy use.

In that respect, it should be noted here that in the most recent FAO outlook (Bruinsma, 2002) the production of biomass other than for food has not been addressed whatsoever. Recently however, the OECD called for "policy changes to promote biomass" (OECD, 2004b). Even though one might argue as to how quickly the rise of a biomass-for-energy sector can evolve, it is nearly impossible to envisage a more sustainable world energy supply without any role for biomass at all.

A large benefit not directly stemming from the increase in available agricultural land is a decrease in pressure on scarce water resources. Meat production is a very large water user world-wide. Millstone and Lang estimate the water use of beef production at 250 m^3 per kg and furthermore state that 1 kg of beef requires one thousand times as much water as 1 kg of cereal (Millstone and Lang, 2003). In their recent publication "Water – More Nutrition per Drop", The Stockholm International Water Institute estimates that the production of 1 kg of grain-fed beef requires 5-40 times as much water as 1 kg of cereal, whilst admitting that their estimate for the water use of beef is rather low (SIWI-IWMI, 2004). In the same publication they describe the current water use in food production as "not environmentally sustainable" and "undermining its own resource base and threatening the resilience of ecosystems".

An extreme scenario as presented here would have enormous impacts on the anticipated growth of the meat sector, as it should have. The impacts are the greatest for the intensive meat producing sectors. One cannot continue the current intensive ways of producing meat if the production of feed crops is discontinued. It should be pointed out, however, that the whole area currently used as permanent pastures is left as is. This means that on a smaller scale than before, cattle farming will still exist and beef and mutton would continue to be available albeit in smaller quantities. The picture for pigs and poultry looks quite different as pigs and poultry are almost exclusively produced intensively and are not kept on pastures. Some pig farming using residues from the food industry as feed might be conceivable, but on the whole these sectors would face a dramatic decrease. It must be noted that such developments would change the whole meat sector worldwide and would require a very powerful driving force as well as a long timescale. Consumers in the current market are unlikely to suddenly stop buying meat at the scale described and so producers are unlikely to stop producing it. Whether consumers in the future will – voluntarily, through price effects or for any other reason – adopt consumption patterns that will lead to reductions in meat production of any significance remains to be seen.

In contrast, it is difficult to see how a significantly large area could be made available for biomass production without the proposed reduction in feed crop production. It is questionable whether much of the land currently used as pastures could be used for growing biomass. Quite often pastures are in areas that are too steep, too cold, too dry, without the necessary infrastructure, etc. to use them for anything but pasture. Using areas that are currently already forested areas seems even more outrageous as the relatively small area of managed forest has its own uses (mainly paper and construction wood) but the not-managed forests and woodlands are often of enormous importance for the preservation of nature (such as tropical forests).

6.3.4 Conclusions

This section clearly shows the interrelationships between food, feed, raw materials and energy from crops. From the calculations presented it is clear that in case of a large-scale transition from meat protein to plant protein the decrease in land needed for feed crops is the largest change in world agriculture. Not only does this change offset any increases in protein production for NPFs, but it also provides an enormous amount of land that can be used for any purpose, as it currently is high-quality cropland.

Our simplified calculations show a biomass potential of one fourth of the current energy use without the detrimental effects that such an increase of biomass use would have if no concurrent reduction in meat production were to take place. There are many benefits conceivably related to a change such as we have presented, in terms of freshwater use, acidification, energy gained (from reductions in fertiliser production, transport etc.) and possibly health effects. Although further study on such benefits is required they are a first but clear indication of how the transitions to more sustainable food, water and energy production – hitherto studied in separation – seem to be inextricably intertwined.

6.4   ACTOR COMMITMENT[4]

Chapter 5 divided the actors who are important for the prospects of NPFs into two main categories, namely (1) the "proximate" decision makers in industry and government, and (2) the members of society who participate in the market and in the political processes. This distinction is highly relevant for the question of how the future of NPFs will be decided when important related issues are on the agenda. The answer is not only dependent on the content but also the timing of decisions. Based on these distinctions, this section will discuss how the initiators of a new technology may gain the commitment of other actors.

Both companies and consumers are sometimes depicted as being "conservative". In view of the meaning of such a term, however, it should be added that it is usually not very wise for companies or consumers to change the course of their behaviour too easily or too often. At the level of an individual company or consumer, the process of behavioural adaptation is not gradual and continuous, as often argued in the innovation literature, but instead it is highly discontinuous (Tyre and Orlikowski, 1994). Hence, for a decision to change an important procedure or habit there are in fact only relatively brief windows of opportunity (see Chapter 1). The conditions that may contribute to the opening of a window are, in general terms, the following:
- implications of another change (e.g. a change in personal relations),

---

[4] Joop de Boer

- aftermath of events that function as interruptions (e.g. food scares, special offers),
- a threshold of gradually increased dissatisfaction (e.g. growing distrust).

These conditions are particularly relevant when initiators of a new technology want to induce changes in the behaviour of companies or consumers. Section 4.3, it will be recalled, focused on the replacement of meat by meat substitutes and specified a number of successive stages in consumer behaviour, which are variants of these conditions. As mentioned in Chapter 1, the multiple stream model of organizational decision making has demonstrated the importance of distinguishing between factors that contribute to the opening of a window of opportunity and factors that increase the chances that the open window will be used in a particular way (Kingdon, 1984). For example, a food scare may temporarily create a window of opportunity for a decision to replace meat by an alternative, but this motivation may not be enough for a lasting behavioural change.

Similarly, the prospects of NPFs may partly depend on the degree to which decision makers in industry and government see them at a certain moment as a solution to sustainability issues, such as the reduction of ecological impacts and as the need to destine more cropland to grow biomass for energy production. As a rule, their decisions on these topics are to a certain extent steered by their anticipations of consumer response in the market and citizen response in the political process. This applies in particular to industries where environmental performance seems to play an important role in the public's perception, such as food and retail. In these cases, there may be strong pressures on government officials to tighten regulations and on companies to do more than what is formally required.

Business strategies

Whether entrepreneurs are interested in the ecological advantages of NPFs depends in part on their specific market. Those companies that are competing in a cost-driven commodity market with largely undifferentiated products, such as oil, grain and meat, will enjoy few financial incentives for achieving environmental or moral performance beyond compliance (Miles and Covin, 2000). In these markets, the price may be the primary marketing variable that differentiates suppliers. They will only be interested in NPFs as far as there are options to produce cheap proteins for multiple purposes.

Other companies may seek to take advantage of the pressure on them and their competitors by incorporating ecological issues into their product or process improvements. The incentives for them to do so may include strategic advantages over their competitors, cost savings, or price premiums for higher quality products. This will require that entrepreneurs can legitimise the inclusion of ecological issues as an integral aspect of corporate identity and that they have enough resources ("discretionary slack") for innovation (Reinhardt, 1998; Sharma, 2000). Generally, this is more likely when price is not the primary basis of competition, when differences between products in a product class are perceived to be significant, when companies are forced to continually improve their products, and when innovations are defensible against imitation by competitors.

Companies that are interested in the ecological advantage of NPFs may use this issue as a selling point in their marketing strategy. Accordingly, they have to choose special signals to gain attention from quality-sensitive customers. This can be done, for example, by bundling information on sustainability issues with product quality information (De Boer, 2003). The resulting quality signals can be transmitted in many forms. Within information economics, it is assumed that the unobservable quality of certain products, such as durables, can be signalled in the form of high prices when customers understand that it is in the economic self-interest of the company to honour its claims about quality (Kirmani and Rao, 2000).

The literature on business strategy and the environment (Peattie, 2001) shows, however, that companies in well-established markets may be reluctant to highlight the relative benefits of a more sustainable product, because they themselves often produce and sell also the conventional brands for which full disclosure would be potentially disadvantageous. Companies may even collectively decide not to compete with each other on a sustainability issue to protect their industry's image and avoid additional costs. Whether the initiators of a new technology will gain the commitment of companies, therefore, depends on the pressure of other actors who may emphasize the relevance of the issue.

Market power

One of the ways in which members of society can influence the decisions of an entrepreneur is by using their market power. Through their participation in the market consumers can specify with their spending that they want a particular type of products. By accepting certain products and rejecting others, consumers can sometimes take sides in moral or political conflicts. In general, however, they do not have much market power over the production process. As Lindblom (2001) notices, their behaviour may affect proximate decisions on what is to be made, but generally not on where and how.

In the case of food, there are still links between types of product and methods of production. Some well-known examples of consumer influence on the ups and downs of production methods involve organic agriculture, genetic modification, and issues of animal welfare. However, market research has shown large differences between consumers in the strength of

their motivation to include, for example, ecological or moral considerations into their purchasing decisions (Roberts, 1996). The following examples show that consumers are often dealing with mixed motives, which may or may not be consistent:
- Consumers who buy foods produced in an ecologically sound manner may primarily be motivated by considerations related to their personal health, which happen to be consistent with ecological considerations (Wandel and Bugge, 1997).
- Consumers who are well aware of the ethical nature of purchase decisions may not change their buying pattern as long as that would be inconsistent with their loyalty to a particular taste, brand or supplier (Newholm, 2000).

Consumers are often accused of paying less attention to ecological or moral criteria when they are in a shop than when they are in a non-commercial situation, but these examples indicate that it is unrealistic to expect that they will simply signal all their preferences through their purchases in the market. As participants in the market, consumers will not ordinarily be focussed on ecological or moral issues and they are often not informed about processing and production methods.

Moreover, if the only merits of a product seem to be that it is considered preferable from an ecological or moral point of view, many consumers might not be fully convinced that they should search for that product and pay a premium price for it. In order to create more value for these consumers, both the design and the marketing of a product should be addressed to all the product attributes that they consider relevant, such as functional and esthetical features, together with distinctive environmental and moral advantages (Meyer, 2001; Peattie, 2001). Depending on the product category (e.g. luxuries or necessities) and the market segment the product is aimed at, this strategy might imply that the product's environmental and moral advantage is presented as one of its self-evident qualities rather than as its main selling point.

In other words, the claim that NPFs have an ecological and moral advantage compared to meat should be only one of its merits in the perception of consumers. As has been said in Chapter 4, consumer-driven product development should not be based on the assumption that consumers will like a product just because of its ecological or moral advantage. Moreover, it is important that this advantage will not be destroyed by any associations between NPFs and controversial processing and production methods. Section 5.2 noted that the wish for more natural ways of food production kept returning with the progress of industrialisation. Those consumers, in particular, who are highly motivated to include ecological or moral considerations into their purchasing decisions may also be highly motivated to scrutinize the naturalness of processing and production methods. Therefore, any ecological and moral claim should be consistent with the overall characteristics of the product.

Issue linking

Another way in which members of society may influence proximate decisions is through their participation in political processes, such as voting or supporting non-governmental organizations (NGOs). In this context, they may be more inclined to take ecological and moral considerations into account than through their participation in the market. Although specific issues, such as climate change and biodiversity, may rise to favour or fall from grace from time to time, their underlying themes will often come back, just as the wish for more natural foods (Lindblom, 2001). This enables individuals and groups to become "political entrepreneurs" who invest resources with the aim to couple a particular solution to a problem. Their activities, as well as factors such as public opinion and public campaigns, may result in the opening of a policy window.

For the opening of a window in the case of NPFs it will matter whether this type of products can be linked with issues such as mentioned in Table 6-4. These linkages (and maybe others) can make it attractive for decision makers in industry and government to consider NPFs as a potential solution to one or more problems on which they have to make decisions. The open window creates opportunities for action inspired by initiatives inside and possibly also outside the organization.

Whether and how the window will be used depends on the position of NPFs on the shortlist of the specialists involved. What the specialists need at such a moment is a viable alternative that they can offer to policymakers. The alternative should be technically feasible and acceptable in terms of relevant values. If the alternative has not been worked out yet, the chances of success will decrease. Table 6-4 gives an indication of the linkages with other issues that may increase the chances of NPFs as cheap proteins and as quality products, respectively. As noted in Chapter 4, the appealing arguments for consumers to reduce meat consumption will be different for various (niche) consumer segments. Notably, the table is just the result of opinions, but it is a tool that can be elaborated in the future.

For instance, the notion that NPFs are essentially plant based may be attractive from various points of view. This notion may appeal to many consumers who have ambivalent feelings towards meat or towards novel meat products, such as meat with a functional property. A recent Canadian study (West *et al.*, 2002) showed that many consumers appeared willing to purchase and to pay a price premium for functional foods, particularly if the functional property, such as anti-cancer substances, were added to foods derived from plants. Consumers were less receptive to a functional property incorporated in a meat product. Although these examples are just hypothetical, they indicate that consumers may have more flexible notions about qualities of plant-based foods than about those of animal-based foods.

Table 6-4. Potential impacts of various issues on the opening of policy windows and the prospects of NPFs as cheap bulk proteins or high-quality specialty products.

Linkages of NPFs with other issues	Contribution to window opening	NPFs' prospects as cheap proteins	NPFs' prospects as quality products
Sustainable food supply at national level	+		
Sustainable food supply at European and global level	+		
North-South issues	+		
Landscape protection	+		
Promotion of biomass	+	+	
Resistance against the treatment of animals	+	+	+
Increasing meat prices	+	+	+
Leaning towards ready meals		+	
Resistance against genetic modification			+
Food safety			+
Prevention of human illness/promotion of human health (e.g. functional foods)			+

In sum, a company's decision on NPFs will be governed by strategic and political circumstances, such as the ripeness of certain issues, at the time the options are contemplated. These circumstances, in turn, will generally depend on its own capabilities, its position in the industry in which it competes, the economic situation of this industry and the industry's public image. Whether an issue is ripe will be influenced, on the one hand, by technological innovations related to sustainability ideals and, on the other hand, by public campaigns that emphasize the ills of an industry. In this relation, the role of NGOs should not be underestimated.

Several issues have been mentioned that may influence the opening of policy windows and the prospects of NPFs as cheap bulk proteins or high-quality specialty products. The main message of this section is that the linkages that may facilitate opening a window of opportunity are often different from the linkages that may improve the market success of a particular product segment (specialty or bulk).

6.5   ACTOR FEEDBACK[5]

6.5.1   Introduction

In the development, market introduction, promotion and support of NPFs a number of different groups of actors are involved. They comprise not only scientific and industrial organisations, but also governmental and non-governmental organisations. Feedback from these actors is indispensable in order to be able to broaden and strengthen the societal basis for the introduction of NPFs and, if required, adjust the research to emerging societal and technological issues. Furthermore, feedback is required for identifying topics missing in the present programme or missing links to other societal issues. The latter is indispensable for developing a follow-up programme.

In order to get feedback, representatives of groups of actors have been interviewed. In these interviews these representatives were asked to give their opinions on enhancing the sustainability of food systems, in particular the meat chain. In addition, they were confronted with the aims and (preliminary) results of the PROFETAS programme and the upcoming ideas for a successor programme. As a start, interviews have been held with policymakers from two groups of actors, i.e. Dutch governmental and non-governmental organisations. The governmental organisations comprised the Ministry of Agriculture, Nature and Food Quality, the Ministry of Economic Affairs and the Ministry of Spatial Planning, Housing and the Environment. In addition, Prof. R. Rabbinge, a Dutch senator, has been interviewed. The non-governmental organisations comprised Friends of the Earth (in Dutch: Milieudefensie), Society of Nature and Environment (in Dutch: Natuur & Milieu), World Wide Fund for Nature (WWF) and the Dutch Consumer Union (in Dutch: Consumentenbond).

6.5.2   Necessity of reducing meat consumption

Sustainability of food systems and developments to improve their sustainability are regarded to be important issues by all organisations. More specifically, the high environmental impact of present meat supply systems

---
[5] Johan M. Vereijken & Harry Aiking

is acknowledged. This impact is thought to be due to the high (indirect) use of resources such as energy, land and freshwater, and the use of polluting chemicals such as pesticides and fertilisers. Taking into account the finite character of the resources, the transition from meat to plant proteins is seen as being inevitable.

In addition to reducing the environmental impacts of meat production, it is expected that securing global future protein supply requires a reduction of meat consumption. However, the Dutch governmental organisations, in particular the Ministry of Agriculture, Nature and Food Quality, stress the point that in the Netherlands neither the per capita meat consumption is rising, nor the number of inhabitants is growing strongly. So in the Netherlands, and probably also in the rest of the European Union, future protein supply is not a high priority item. However, in other parts of the world, especially in rapidly industrialising countries such as China and India rising meat consumption (due to an increase in both consumption per capita and in number of inhabitants) will probably result in a global protein shortage. Of course, this will affect world-wide protein supply chains. In particular, this will create problems for protein supply in developing countries. For instance, in these countries the pressure to produce protein for global use at the expense of production for regional or local use may increase due to rising global protein prices. Therefore, problems associated with increasing global meat consumption clearly have a North-South dimension.

Furthermore, a reduction in global protein supply may result in changes in flows of proteins to Europe. For instance, to feed its animals, Europe is a major importer of soy protein from countries such as the USA and Brazil. Rising prices of soy proteins may therefore result in a decrease of this import. To cope with this, the European production of protein-rich crops, which is relatively low at this moment, may increase. The European policy is still directed towards self-sufficiency. However, some policymakers expect that this will become less and less important because of the globalisation of feed and food supply. Based on this train of thought, neither the Netherlands nor the European Union is seen as being the "problem owner" with respect to securing protein supply. However, the Ministries, and in this case particularly the Ministry of Agriculture, Nature and Food Quality, assume they have a role in raising the awareness that a problem is coming up.

6.5.3 Ways to reduce the environmental impact of meat consumption

To reduce the environmental impact of meat production, European non-governmental organisations generally promote the consumption of organic meat or simply a reduction in meat consumption. The latter could be enhanced by higher prices for meat, but it is generally felt that just taxing meat is not the way to go. In their opinion, a better way would be internalising environmental costs into the price of meat (see also the arguments presented in Section 5.4).

A technological way to reduce the environmental impact of meat production is by so-called precision farming: during cultivation of fodder crops resources such as nutrients, water and pesticides are used only as required in a particular spot of the field, in the right amounts and on the right parts of the plants. The sustainability of meat produced by this kind of farming is expected to be better than that of organic meat. Other technological ways that have been mentioned are feedback coupling in chains (e.g. just-in-time production) or by lateral integration between different production chains (coupling of chains in such a way that by-products from one chain are available at the right moment and amount for use as inputs in another chain). Of course, another technological option is the incorporation of plant proteins in meat products.

With respect to the PROFETAS programme all interviewees agreed that the introduction of NPFs, provided that they are successful among consumers, will result in reducing meat consumption and, hence, in the associated environmental impact.

6.5.4 Barriers

Technological and societal

The main bottlenecks in introducing NPFs are not considered to be technological. The (non-technological) actors expect that technology can provide the necessary tools to develop NPFs that meet consumer demands. However, it is generally agreed that present meat substitutes – though on the rise – are not very successful in the market and, hence, not in reducing meat consumption. Other products that better agree with present trends should be developed. For instance, products that align with trends towards so-called "grazing" (more eating moments a day), towards exotic eating habits (eating of foreign foods) and towards healthy products (reduction of risk for diseases).

According to all organisations the main bottleneck is the consumer's attitude. They regard the consumer to be "conservative" with respect to willingness to change consumption behaviour. Furthermore, all stressed that in general the consumer does not want to pay for more environmentally friendly products though the consumer says so (the citizen – consumer dilemma). However, times may be changing. For instance, consumer concerns for animal welfare are increasing and may be one of the driving forces to reduce meat consumption.

Legislative

Most organisations expect that new and previously underestimated legislative rules might hamper the introduction of NPFs. Specifically it was mentioned that the Novel Food Regulation of the European Union could be an obstacle, because it is a rather expensive and time-consuming procedure. Furthermore, specific agreements regarding land use and agricultural produce in the framework of the World Trade Organisation and/or the European Union might constitute obstacles. The Ministries agreed that the government should ease legislative obstacles in order to facilitate the development of NPFs. It is suggested that a way of doing this might be to direct legislation more towards goals and less towards means or instruments.

6.5.5 Opposition

In the opinion of the actors, opposition against the introduction of NPFs might come from the part of the agro-lobby that is involved in meat production. In particular, the Netherlands is an important meat producing and exporting country. So, in the short term opposition could be large in the Netherlands, but in the long term meat producers might get involved in the production of NPFs. Furthermore, the Netherlands is an important producer of food and home country to many large food companies. So the Netherlands might even take the lead in introducing NPFs.

6.5.6 Introducing Novel Protein Foods into the market

Consumer aspects

All organisations agreed that NPFs can significantly contribute to the sustainability and security of future protein supply. However, to achieve a large consumption will be difficult because, as already pointed out, consumers are "conservative" and not considered willing to abandon meat consumption. To be successful, NPFs should not be too expensive compared to meat. The Dutch Consumer Union stressed that consumers do seem to accept higher prices for increased animal welfare and sustainability, provided they know what they pay for. However, the price difference may not be too large and clear information about the background should be available to the consumer.

In addition to sustainability, health aspects may be an issue. For instance, the fat content and composition (saturated versus unsaturated fat) can be adjusted much more easily in man-made products such as NPFs than in meat. This kind of health aspects may enhance the acceptability of NPFs by the consumer. In addition, consumers are willing to pay for specific health claims as is evident from the price difference between low fat spreads with and without added plant sterols. When informed that NPFs are overall healthier than meat this might also enhance consumer acceptability.

Initiative

According to most organisations, the initiative to introduce NPFs on the market should come from industry, particularly large food companies. The latter have the marketing facilities, know-how and power to introduce real new food concepts such as NPFs. This is required because according to the interviewees the consumer is, as already mentioned, "conservative" with respect to food choice and therefore has to be tempted to buy NPFs. To this end, these products should not be marketed as meat substitutes but given a unique image. They should not be directed towards vegetarians or to other idealist groups. Instead "they should have a trendy image". In line with this reasoning, most organisations feel that NPFs should initially be marketed as a speciality. However, to have a substantial effect on either or both the environment and protein supply, it is agreed that NPFs should become a commodity.

Support

It is generally stated that the government should play a role in the introduction of NPFs by facilitating their introduction. Not by subsidising them, but by other means. A covenant covering the whole chain from primary producers to retailers could be helpful in this respect. In addition, support may come from non-governmental organisations. However, care should be taken, because some consumers associate some of these organisations with activists.

6.5.7 Relation to other transitions

Governmental initiatives

In the fourth National Environmental Policy Plan four transition programmes are outlined. The transition from meat to plant protein consumption by introducing NPFs, the focus of the PROFETAS programme, is regarded to have links to three of these four transition programmes:
- Sustainable agriculture
- Biodiversity
- Energy

The link to sustainable agriculture is obvious: a sustainable cultivation of protein-rich crops, the starting material of NPFs, will contribute to enhancing the sustainability of NPFs. Furthermore, the environmental impact of sustainable agriculture will be decreased by the reduction of meat production in favour of the more environmentally friendly NPF production.

The link to biodiversity is also clear-cut: rising meat consumption threatens biodiversity, because it will result in higher usage of natural resources such as land and freshwater. Furthermore, rising meat production threatens biodiversity by the resulting increased levels of air, water and soil pollution.

One of the topics within the transition programme "Energy" is the use of biomass to produce energy in a sustainable way. The production of biomass implies the production of proteins, a component too valuable to convert into energy. Usage of this protein to produce NPFs will contribute to enhancement of the economics of biomass utilisation. Vice versa, the biomass produced by the production of plant proteins can be used to produce energy and in this way contribute to the overall economy of NPF production.

Non-governmental initiatives

The World Wide Fund for Nature has recently started projects directed at the reduction of freshwater usage. Because meat production requires much more water than plant protein production does, the relation to the PROFETAS programme is evident.

6.5.8 Conclusions

In line with the PROFETAS programme, all organisations state that the environmental impact of the present meat supply is high and should be reduced. Furthermore, it is agreed that rising meat consumption is most likely to result in a shortage in future protein supply. However, in the Netherlands reduction of meat consumption does not have a high priority yet. The ministries stress that with respect to future protein supply, neither the Netherlands nor the European Union is the problem owner.

The organisations indicated that there are several ways to reduce meat consumption. The introduction of NPFs will, provided they are successful among consumers, certainly contribute to this reduction. However, this introduction will be difficult because the consumer is felt to be "conservative" with respect to food choice. In order to deal with this obstacle the initiative to introduce NPFs should come from large companies because they have the means to tempt the consumer. In line with this argument, most organisations feel that NPFs should initially be marketed as a specialty.

With respect to the follow-up programme of PROFETAS, most actors advised placing the transition of meat to plant proteins in a global context and addressing the North-South dimension. Furthermore, links should be established to other transition programmes.

In conclusion, the organisations agree with the aims and ideas of the PROFETAS programme. However, developing NPFs to be consumed on a large scale is not an easy task. In a follow-up programme more emphasis should be on the global context of the meat to plant protein transition as well as on links to the other three transitions.

6.6 CONCLUSIONS [6]

The results of the PROFETAS programme should be seen in the context of an evolving global food economy, which will increasingly recognize the advantages of crop-based solutions. For various reasons, decision makers in industry are exploring options for a cross-fertilisation between food disciplines and areas such as biotechnology and information technology. Generally, large companies seek greater economies of scale for global and highly flexible supply chains. New technologies will increase the possibilities for product differentiation and improve responsiveness to customer demand. Clearly, consumers and society in general will influence the direction of these developments, but there are several major areas of uncertainty in which the future of the sector remains open. In many countries, for example, there is uncertainty about the public's willingness to pay for environmental quality and about the possibility that specific areas of technology will be implemented and accepted.

Protein-rich foods are playing an essential part in the pursuit of a food economy that contributes to sustainability and health. The results of PROFETAS clearly indicate that a societal transition from meat to plant protein is indispensable towards the achievement of sustainability. However, how such a transition may be realised – in particular with regard to social and technological aspects – is yet unclear. A requisite, but in itself insufficient condition for this leap towards a more sustainable food production is to convince the consumer, as can be seen from Section 6.4. Since Western European consumers tend to be more susceptible to arguments regarding their health than to sustainability arguments, the more feasible approach seems to be human health, targeting issues such as obesity, circulatory disease and food safety. In other countries, however, it may be crucial to link a diet shift with culinary traditions.

[6] Harry Aiking, Joop de Boer & Johan M. Vereijken

Due to the approach of global sustainability from a European perspective, the initial PROFETAS focus has been on pea as a model crop. In Western Europe, however, human consumption of pulses has been declining for centuries (Smil, 2001: 36). Globally, soy is the most abundant pulse crop, but due to climatic requirements, cultivation of soy is taking place almost exclusively in the Americas and in Asia. The main product is oil for human consumption. 95% of the remaining cakes, which are high in protein, is primarily used in livestock feed. A small, but culturally and historically significant, amount of soy is used to produce protein-rich products such as tofu, tempeh and miso, which are primarily consumed in Asia.

Climate and soil determine to a large extent which crops can be grown efficiently and where, as can be seen from our crop growth modelling research in Section 3.4. Whether pea-derived NPFs will be the products of choice for a transition in Europe remains to be seen. Alternatively, it seems likely that soy-derived NPFs may be a successful option in Asia. Not only would this option also abolish much transportation, but also it is likely that soy would appeal more to Asian consumers than peas, due to its long-standing historical and cultural familiarity. The Americas may constitute an intermediate case. At any rate, it is evident that local production and consumption coupling has huge environmental advantages due to transport reduction. Consequently, in different parts of the world different crops are likely to be used for NPFs production.

A partial diet shift in Western Europe and other OECD countries may function as a social and technological model for producers and consumers in other parts of the world. This applies to both the "eat less meat" option as the "eat more plant-based protein" option. One of the notions behind PROFETAS was that scientific innovation might take place in Europe, for example, and then diffuse to other regions (Herok, 2003). Although Europe, and particularly north-western Europe, has often been characterised as a meat eating continent, such a transition is not altogether unlikely. Currently a trend towards a more vegetarian lifestyle seems to be slowly emerging here, which seems to have been shaped by an endless string of animal-related food safety incidents. These incidents have upset Western consumers in particular because health-consciousness and protection of animal welfare are on the rise. Interestingly, the concept of sustainability itself has not emerged as a key issue valued by many Western consumers, as it may be too remote from their daily lives.

In contrast, in developing countries food security (Gupta, 2004) is valued over both food safety and sustainability. Furthermore, it is in the rapidly industrialising countries, in particular, where most of the world's population and/or economic growth is taking place. In China, for example, a moderate population growth is coupled with a high economic growth and, thus, leading to a booming meat demand. On the one hand, China is experiencing meat production shortages and, on the other hand, massive pollution in the central provinces, such as Sichuan, in which most of the production is concentrated (Sicular, 1985; Ke Binsheng, 2005). Geissler (1999) argued that in China a (nutrition) transition is presently taking place from soy to pork rather than the other way around and, in this respect, she concluded that in China "nutritional policies to promote the consumption of soyabean are unlikely to be effective in the context of an increasingly free and global market." India is a case in itself (1) because its rapidly expanding population is expected to outnumber the Chinese within the next few decades and (2) due to its vegetarian background, which is already showing some cracks, maybe as a result of increasing affluence. In summary, it is clear that regional approaches and intercontinental cooperation will be indispensable to achieve a worldwide protein transition towards sustainability.

REFERENCES

Berk, Z. (1992), *Technology of production of edible flours and protein products from soybeans,* Technion, Israel Institute of Technology, Haifa.

Bruinsma, J. (2002), *World agriculture: Towards 2015/2030, An FAO perspective,* FAO, Rome.

De Boer, J. (2003), "Sustainability labelling schemes: The logic of their claims and their functions for stakeholders", *Business Strategy and the Environment,* Vol. 12, pp. 254-264.

De Haan, C., Steinfeld, H., and Blackburn, H. (1997), *Livestock and the environment: Finding a balance,* European Commission, Directorate-General for Development, Brussels, Belgium.

Dijkstra, D.S., Linnemann, A.R., and Van Boekel, M.A.J.S. (2003), "Towards sustainable production of protein-rich foods: Appraisal of eight crops for Western Europe. - PART II. - Analysis of the technological aspects of the production chain", *Critical Reviews in Food Science and Nutrition,* Vol. 43 No. 5, pp. 481-506.

FAO (2004), *Protein sources for the animal feed industry,* FAO, Rome.

FAO (2005), "FAOSTAT data", Available at <faostat.fao.org/default.jsp?language=EN>.

Geissler, C. (1999), "China: The soybean-pork dilemma", *Proceedings of the Nutrition Society,* Vol. 58, pp. 345-353.

Gupta, J. (2004), "Global sustainable food governance and hunger", *British Food Journal,* Vol. 106 No. 5, pp. 406-416.

Herok, C.A. (2003), *The introduction of meat substitutes from vegetable proteins: Consequences for EU agriculture on a global level,* PROFETAS Project I3B Final Report, LEI, The Hague.

Hoogwijk, M., Faaij, A., Van den Broek, R., Berndes, G., Gielen, D., and Turkenburg, W. (2003), "Exploration of the ranges of the global potential of biomass for energy", *Biomass and Bioenergy,* Vol. 25 No. 2, pp. 119-133.

Ke Binsheng (2005), "Industrial livestock production, concentrate feed demand and natural resources requirements in China", Available at <faostat.fao.org/WAIRDOCS/LEAD/X6146E/X6146E00.HTM>.

Kingdon, J.W. (1984), *Agendas, alternatives, and public policies,* Harper Collins, New York.

Kinsella, J.E. (1976), "Functional properties of proteins in foods: A survey", *Critical Reviews in Food Science and Nutrition,* Vol. 7, pp. 219-280.

Kirmani, A., and Rao, A.R. (2000), "No pain, no gain: A critical review of the literature on signaling unobservable product quality", *Journal of Marketing,* Vol. 64, pp. 66-79.

Lindblom, C.E. (2001), *The market system: What it is, how it works, and what to make of it,* Yale University Press, New Haven.

Linnemann, A.R., and Dijkstra, D.S. (2002), "Towards sustainable production of protein-rich foods: Appraisal of eight crops for Western Europe. - PART I. - Analysis of the primary links of the production chain", *Critical Reviews in Food Science and Nutrition,* Vol. 42 No. 4, pp. 377-401.

Meyer, A. (2001), "What's in it for the customers? Successfully marketing green clothes", *Business Strategy and the Environment,* Vol. 10, pp. 317-330.

Miles, M.P., and Covin, J.G. (2000), "Environmental marketing: A source of reputational, competitive, and financial advantage", *Journal of Business Ethics,* Vol. 23, pp. 299-311.

Millstone, E., and Lang, T. (2003), *The atlas of food: Who eats what, where and why,* Earthscan Publications Ltd, London.

Newholm, T. (2000), "Consumer exit, voice, and loyalty: Indicative, legitimation, and regulatory role in agricultural and food ethics", *Journal of Agricultural and Environmental Ethics,* Vol. 12, pp. 153-164.

OECD (2004a), *Agricultural outlook 2004-2013,* Available at <www.oecd.org>, Organisation for Economic Co-operation and Development, Paris.

OECD (2004b), *Biomass and agriculture: Sustainability, markets and policies,* Organisation for Economic Co-operation and Development, Paris.

Peattie, K. (2001), "Golden goose or wild goose? The hunt for the green consumer", *Business Strategy and the Environment,* Vol. 10, pp. 187-199.

Reinhardt, F.L. (1998), "Environmental product differentiation: Implications for corporate strategy", *California Management Review,* Vol. 40, pp. 43-73.

Roberts, J.A. (1996), "Green consumers in the 1990s: Profile and implications for advertising", *Journal of Business Research,* Vol. 36, pp. 217-231.

Sharma, S. (2000), "Managerial interpretations and organizational context as predictors of corporate choice of environmental strategy", *Academy of Management Journal,* Vol. 43, pp. 681-697.

Sicular, T. (1985), "China's grain and meat economy: Recent developments and implications for trade", *American Journal of Agricultural Economics,* Vol. 67 No. 5, pp. 1055-1062.

SIWI-IWMI (2004), *Water - More Nutrition per Drop,* Stockholm International Water Institute, Stockholm.

Smil, V. (2001), *Enriching the earth: Fritz Haber, Carl Bosch, and the transformation of world food production,* MIT Press, Cambridge (Mass.).

Smil, V. (2002), "Worldwide transformation of diets, burdens of meat production and opportunities for novel food proteins", *Enzyme and Microbial Technology,* Vol. 30, pp. 305-311.

Tyler, R.T., Youngs, C.G., and Sosulsky, F.W. (1981), "Air classification of Legumes. I. Separation efficiency, yield and composition of the starch and protein fractions", *Cereal Chemistry,* Vol. 58 No. 2, pp. 144-148.

Tyre, M.J., and Orlikowski, W.J. (1994), "Windows of opportunity: Temporal patterns of technological adaptation in organizations", *Organization Science,* Vol. 5, pp. 98-118.

Van den Broek, R. (2000), *Sustainability of biomass electricity systems,* Eburon, Delft, The Netherlands.

Wandel, M., and Bugge, A. (1997), "Environmental concern in consumer evaluation of food quality", *Food Quality and Preference*, Vol. 8, pp. 19-26.

West, G.E., Gendron, C., Larue, B., and Lambert, R. (2002), "Consumers' valuation of functional properties of foods: Results from a Canada-wide survey", *Canadian Journal Of Agricultural Economics*, Vol. 50, pp. 541-558.

CHAPTER 7

TRANSITION FEASIBILITY AND IMPLICATIONS FOR STAKEHOLDERS

Harry Aiking, Joop de Boer & Johan M. Vereijken

7.1 INTRODUCTION

Taken together, the research projects that constitute PROFETAS have developed a comprehensive approach to assess whether and how a transition of protein production and consumption may contribute to a more sustainable system of protein supply. The common object of study was the triple hypothesis that a shift in the Western diet from meat protein to plant proteins is (1) environmentally more sustainable than present trends, (2) technologically feasible, and (3) socially desirable. In the design and evaluation of alternative protein chains the focus was on NPFs and consumer preferences were given a predominant role. For the comparative analyses, the entire protein chain served as a starting point to measure differences between the NPF sector and the intensive livestock sector in the Netherlands and the EU. Given the global nature of the concept of sustainability, environmental, economic and social issues were studied in the context of global development. In addition, several emerging options were pointed out from the general perspective of a future biobased economy.

No matter how comprehensive, studies of this sort inevitably show strengths and weaknesses. One of the PROFETAS' strong points is the toolbox principle it adopted. According to this principle, developing the methodology and the tools now to facilitate problem solving in the future is far superior to trying to solve presently perceived future problems. Another strong point is the programme's multidisciplinary character and its embedding in more than one university. This multifaceted setting paved the way for a flexible organizational structure with various linkages to representatives from science, government, industry and, to a lesser degree, NGOs. From the very beginning, the organization was especially designed to ensure outcomes that may assist policymakers from both government and industry with relevant tools and arguments.

An evident weakness is the programme's dependency on pragmatic choices, boundary conditions and other explicit assumptions. The main cause is that whereas the meat chain has been optimised for hundreds of years, the plant-based NPF chain is still in its infancy. Hence, the initial choices were made to focus the projects and to get them started. Methodologically, PROFETAS is very much a product of systems analysis, a tool that approaches complex issues, such as sustainable food production, by constantly making trade-offs between the analyst's workload (reduced by disregarding non-core issues) and the output's realistic portrayal (reduced by each simplification). In fact, it is the most flexible and robust research strategy that can be devised in this type of studies.

Surprisingly, some of the most remarkable insights PROFETAS delivered refer to a number of choices from the programme's very beginning, which have been put in a different light. For instance, it was decided at the outset to study the global food system from a national (Dutch) and European perspective. This seemed a logical choice, because – as far as we are aware – the Netherlands is the only country in the world with an explicit policy on "transition management" towards improved sustainability. An implicit assumption may also have been that developed countries are setting the global consumption standard. Although the latter may still be true at present, it has become clear that in rapidly developing countries such as China and Brazil the incentives to achieve a protein transition are not just different, but also likely to be much stronger. As a consequence, it may become more fruitful to look from the international context towards the EU and the Netherlands than the other way around.

In other words, it may primarily be global environmental and societal changes that contribute to the opening of a policy window for a diet shift, both in the Netherlands and elsewhere. However, whether the open window will be used in a particular way, may also depend on decisions by policymakers from government and industry at the national level. Although the Netherlands as a country is far from representative with respect to agriculture and with respect to freshwater availability, it may significantly contribute to the development of alternative protein chains. Against this background, the rest of this chapter is devoted to discuss the insights and tools PROFETAS delivered, as well as related options and arguments.

7.2 ENVIRONMENTAL SUSTAINABILITY

7.2.1 Insights and tools delivered by PROFETAS

In a highly simplified form, a production system may be called sustainable if it can be maintained at a certain "quality" in the long run. The quality may be evaluated by monitoring trends in a number of indicators, which are supposed to reflect complex and sometimes poorly understood phenomena. Thus, in order to describe environmental sustainability in relation to alternative protein chains, in PROFETAS three approaches were chosen. This multi-method strategy is based on the notion that the mutual relationships of their findings may shed light on different aspects of a complex phenomenon. In theory, these findings may either converge and validate one another, or they may highlight complementary aspects of the phenomenon, or they may show scientifically interesting divergences. In each of the approaches, two model protein chains (i.e. the pork chain and the pea-based NPF chain) were used to analyse the consecutive stages associated with a certain amount of human protein consumption.

First, in the *environmental* approach a tool was developed to find a systematic way to select the substances and processes that are relevant for a comparison of pork and pea-derived NPFs. The option chosen is to start off from the reliance of human society on the benefits that the environment provides and, therefore, to focus on ecosystem services. For using ecosystem services as the theoretical basis, it is possible to provide transparent criteria for a comparative study of alternative protein food options. In addition to its more solid theoretical foundation, the method displays – as intended – several advantages over established tools such as LCA, including the focus on effects rather than on emissions, applicability to virtual products and the inclusion of location-dependent parameters such as climate and soil properties.

Second, in the *economic* approach a model linking economy and environment was developed, which yielded the view that the problems related to pork production in the Netherlands and the European Union are not only local but also global, as a result of large-scale imports of feed. It was concluded that for global environmental sustainability a local (European) transition in itself will be insufficient, because the current centres of growing economies plus growing meat consumption are to be found in Asia and South America. Whether it is profitable to continue local pork production for economic reasons is partly dependent on the environmental standards set by governmental policymakers. In this respect, the model parameters illustrated that – under the present circumstances – the political role of citizens who support more stringent environmental standards might be greater than their economic role as consumers when they make trade-offs between the prices of pork and NPFs.

Third, the *ecological* approach argued that for developing a coherent *set* of indicators no framework turned out to be available, which (1) links pressures to environmental consequences, and (2) focuses on how each of the indicators can contribute to the problem solving logic required to answer questions of environmental sustainability. Rather, the existing frameworks take a one-dimensional perspective. Therefore, causal networks were recommended, describing how individual indicators are interrelated, thus providing a more complete, holistic picture of what is happening in the environment. In short, process knowledge provides insight into cause-and-effect networks and is put forward to guide the selection of appropriate indicator sets.

Together, irrespective of the approach, a clear consensus was provided (1) that NPFs are environmentally more sustainable than pork and (2) that the most important impacts are due to tampering with (a) land use and the cycling of (b) water, (c) carbon and (d) nitrogen, leading to biodiversity loss, climate change, eutrophication and acidification (Chapter 2).

7.2.2 Further insights and tools required

Given the convergent results concerning the relative advantages of NPFs, the question emerges whether this information will be enough for policymakers who want to choose between alternative protein chains. As we shall see in greater detail later, policymakers should try to avoid suboptimal solutions and that makes it important for them to start with a comparative analysis of a large number of options. Presently, the environmental assessment tool is limited to the agricultural production phase. After extension to the rest of the production and consumption chain the method should be generalised beyond comparing just protein food options. In view of the increasing relevance of linkages between transitions, it should be extended sufficiently to become a tool to assess the environmental sustainability of transitions in general.

Because a large part of the environmental impact of the pork chain takes place outside Western Europe, it is important to also look at the global dimension. Reducing this environmental impact in developing countries of Asia and South America by shifting to a more plant protein centred consumption pattern in Western Europe may have important economic repercussions in those developing countries. In turn, this may or may not cause negative environmental impacts there, depending on what alternative livelihood strategies will be developed to compensate reduced feed crop exports to Western Europe. So North-South issues clearly require further study.

7.3 TECHNOLOGICAL FEASIBILITY

7.3.1 Insights and tools delivered by PROFETAS

Six projects aimed at assessing aspects of the technological feasibility of pea-derived NPFs were performed in PROFETAS. Two of them were directed at aspects concerning primary production of peas and two at aspects concerning processing of peas to NPFs. In addition, in one project a tool is being developed to optimise the NPF chain. Last but not least, in one project the options for the non-protein fractions were studied.

In the *breeding* project it was shown that between different varieties there is a large variation in protein composition. This variability can be exploited in classical breeding programmes to obtain pea varieties that are optimised towards NPF production. Furthermore, a tool has been developed for genetic modification of pea and other legumes. Its application will enable selective removal of certain storage proteins (which affect texture formation) and of certain enzymes (which affect flavour compounds) in order to improve the quality of the raw material. As an alternative, genetic modification may be used in R&D for screening the effects of alterations in protein and flavour composition. Such an approach will speed up classical breeding while field crops are kept GM free.

In the *cultivation* project, a crop growth model has been developed that predicts yield of peas and pea protein as a function of genotype and environmental variables (such as solar radiation, temperature and rainfall). The model is generic; it can be extended to any arable crop. The model can be used to optimise production (e.g. with respect to resource use efficiency such as water) or to define the characteristics a variety should have for a given environment. An important insight resulting from this project is that using varieties with a high protein content cannot increase the protein yield per hectare. To achieve the latter, other traits (e.g. plant architecture to sustain leaf area duration during the grain filling period) should also be changed. Furthermore, attention was drawn to agronomic aspects, such as resistance to lodging and soil-borne diseases, which indicate that peas may not be the best choice to function as starting material for NPF production.

The *texture formation* studies yielded new insights into a process that is of prime importance in texture formation, i.e. heat-induced gelling behaviour of pea proteins. This behaviour was shown to be affected by the protein composition and by the rate of cooling during processing. This implies that a range of textures can be produced (a) by varying the protein composition by selecting appropriate pea varieties and (b) by varying the processing parameters. This provides basic tools to food technologists to direct the texture of pea-based NPFs towards consumer demands. Comparative studies using soy and pea proteins led to the insight that despite similarities between proteins on the molecular level, their gelling behaviour may be different. On the one hand, this could mean that a large range of textures can be made using a limited set of proteins. On the other hand, this observation likely sets limits to the interchangeability of proteins.

Protein-flavour interactions studies using model volatile flavour compounds provided information on the amounts of flavour compounds associated with the pea proteins under various conditions. They also showed that the interactions with flavour compounds differ per type of pea protein. So again, varying protein composition or process parameters offers possibilities to tune the flavour of NPFs. Furthermore, it was shown that saponins, not only present in peas but also in numerous other legumes (including soy) are responsible for a bitter taste. This bitterness may be reduced by heating, thus providing a technological tool to adjust taste. It was also found that in pea protein preparations "off-flavours" seem to be present that are derived from fat oxidation. Similar "off-flavours" have been found in other plant protein preparations too; very well known examples are soy protein preparations.

The project on *chain design* will not be completed before 2006. Tools developed in this project are expected to be useful in the design of food chains to optimise these chains either towards a single goal (such as product quality, costs, or environmental load), or even towards multiple goals. Preliminary results are that the cost price of NPFs could be less than or at least comparable to that of pork. Comparison of the environmental load by using exergy analysis showed that an NPF chain is more efficient than a pork chain only if the non-protein fractions of pea seeds (mainly starch) are put to use.

Options for the non-protein fractions were investigated in a project added later to study the feasibility of plant-derived NPFs from a different perspective. A tool was developed to assess different protein crops on the potential uses of the non-protein fractions. Only bulk fractions were distinguished and evaluated on their suitability to be used as food, feed, stock and biofuel. The conclusion was drawn that – at present – oil crops may be at an advantage, because the oil fractions have better application perspectives than other non-protein fractions. Interestingly, a transition from intensive meat production to NPF production from any crop would lead to a shortage of soy oil for food applications, rather than to a surplus of this non-protein fraction. If it is assumed that all *intensive* livestock farming disappears and only grass-fed and other extensive meat production would remain, then this shift – irrespective of the crop (pea or soy) – would release an enormous area (over 300 million ha) of highly productive land for other uses. For example, if used to grow biomass, this would be sufficient to cover approximately 25% of the current world energy demand.

7.3.2 Further insights and tools required

The research on technological feasibility has focussed primarily on basic problems and on issues that needed to be solved before products could be designed. With respect to processing some important insights and tools are still missing. For instance, neither have texturisation processes been studied, nor have products (or product concepts) actually been made. Such research is indispensable to allow a consumer-oriented development of NPFs. At present it is still unknown whether more traditional texturing processes such as extrusion can deliver the range of textures desired by consumers. It may well be that novel techniques have to be deployed and further developed, such as techniques based on phase separation. Furthermore, in actual NPFs other components (e.g. hydrocolloids, fats) will be present. No information is available about the effects of such components on the texture and flavour of the NPFs. With respect to flavour, stakeholders from industry have indicated that starting materials for NPF production should preferably be free from "off-flavours". The prevention of formation of such compounds or ways to remove them should be investigated. A lot of knowledge on this subject is already available from research on soy proteins.

Missing insights with respect to the use of the newly developed protocol to genetically modify peas include, among others, its effectiveness and efficiency as well as its effects on cultivation. These effects not only include those relevant for NPF processing (e.g. effects of protein composition on texture and flavour) but also those relevant for primary production. Questions have to be answered regarding the effects of the modification on for example (a) germination power of the seeds, (b) viability of the modified peas in the field and (c) an eventual yield penalty for total protein content because of the modification. Furthermore, the applicability of this new protocol to modify crops other than peas needs to be further explored.

To exploit the interesting possibilities offered by the crop growth model, it has to be validated and to be extended to crops other than peas. Then it can be a powerful tool in identifying differences among various crops in resource use efficiency for e.g. biomass and protein production, hence in contributing to a sustainable primary production of raw materials for NPFs.

In summary, quite a lot of research is still required to actually develop and produce NPFs that meet consumer preferences. In addition to the technological issues discussed above, the project on chain design will most likely result not just in answering questions but also in raising others. Furthermore, options for the non-protein fractions should be detailed by means of scenario studies, with particular attention for the projected shortage of soy oil for food applications. Finally, the issue of the crops to be used for NPF production is still open. As argued in Section 6.2, even in Europe peas may not be the preferred crop for NPF production. More research concerning crop choice is required.

7.4   SOCIAL DESIRABILITY

7.4.1   Insights and tools delivered by PROFETAS

In addition to sustainability – which is an evident societal preference – societal desirability was studied in PROFETAS with respect to (1) the behaviour and preferences of consumers and (2) the food-related political and economic developments of the next decades. The underlying notion of this work was that the desirability of a diet shift is in proportion to its fitting in with the behavioural patterns of current and future generations. In addition, it was argued that a lack of fit would create less serious problems if mitigating measures can be taken.

Based on a long-term view on behaviour, the potential for a diet shift in relation to socio-cultural changes was examined. A newly developed analytical framework sorts insights on influences on behaviour into a logical order. Its cascade-like structure embodies the view that a long-term development will create opportunities for food choices that match its general direction, whereas it will put constraints on others. The analysis of long-term changes indicated that there is a favourable socio-cultural context for decisions that make consumers and producers less dependent on meat proteins. One of the most salient results of this work, however, is the contrast between, on the one hand, a series of impressive changes in dietary choices over the last few centuries and particularly over the last few decades and, on the other hand, the observation that an individual will not easily change his or her food preferences from one day to the next. This contrast underlines the value of the analytical framework in combination with small-scale consumer research.

By using currently available meat substitutes as a model to analyse consumer behaviour and food choices, it was possible to develop several tools that may guide NPF product development. It appears that the currently available meat substitutes will not become popular without additional measures. Analysis of consumers and consumption behaviour with respect to meat and meat substitutes provided insights such as (a) non-vegetarian consumers of meat substitutes are not impressed by environmental arguments, (b) only a small group of consumers is open to completely new products due to so-called "neophobia", (c) consumers would like more information on usage and preparation of meat substitutes and on ingredients used.

Consumer sensory preferences were analysed and their translation into product characteristics was attempted, yielding insights such as (a) attention should be paid to satiating properties of NPFs, which are related to protein content, (b) most people want soft, smooth or crispy meat substitutes to have a seasoned, meat-like flavour and a brown colour, (c) meat substitutes should have the same place in the dish as meat. Other than originally anticipated, the latter results clearly show that people want NPFs to have meat-like characteristics. They confirm the notion that people will habitually look for what is familiar when they are trying to make sense of something, such as an invitation to try a product. The retrospective character of sense making can explain why non-vegetarian consumers keep relying on distinctions drawn in the past and why they evaluate meat substitutes by using meat-based criteria.

The selection environment in which NPFs have to survive is partly dependent on their positioning in the market, for example as cheap (bulk) proteins or as quality products (specialties). The prospects of the new products may also depend on linkages with other issues, such as increasing meat prices or health promotion. These prospects are not only dependent on consumer responses in a potential usage situation, but also on processes at the level of markets and public institutions. Given the fact that technology development takes place in a changing and malleable world, the stakeholders of a new technology may opt to monitor and influence its selection environment. PROFETAS applied some advanced tools such as policy analysis and econometric modelling, as well as a skilled examination of the rules of international institutions with regard to novel foods.

In a political analysis it was investigated how politics and public policy affect the possibilities of a diet shift. As expected, it appeared that governmental policy does not have many direct influences on food choices (i.e. the proportion of meat vs. plant proteins in the nation's diet), but mostly indirect influences. These influences demonstrate that production and consumption are, at least partially, facilitated by a political "infrastructure", implicitly favouring certain products or production processes over others. To uncover indirect influences, it was analysed how current food practices have developed in the Netherlands since 1850 (i.e. when the physician G.J.

Mulder brought up the issue of proteins and the societal effects of the lack of proteins in most people's diets). Given the current sustainability-related issues, the political infrastructure still seems to favour the option of more protein production and consumption, although that option is not actively promoted by governmental action. From a sustainability perspective, however, it may be desirable to have a political infrastructure that favours a more divergent food system with different approaches to food. This would mean, for example, that the option of producing more plant protein does not necessarily get more emphasis than the option of simply "eating less meat".

Two complementary econometric analyses studied agricultural production and consumption and the patterns exhibited (by adaptations to the existing GEMAT model), and institutional arrangements affecting agricultural production and trade (adapting the GTAP model), respectively. It was shown that a protein transition results in a decrease of environmental pressure on land under all scenarios (GTAP) and that the potential is even larger (GEMAT). Though crucially important to the feasibility of a transition, interestingly, the two models employed disagree on future meat price development. From GTAP a price decrease may be inferred, but from GEMAT a price increase. The underlying cause of this disagreement is the implicit assumption, deeply embedded in the models, to what extent agricultural efficiency will continue to increase in the future. These results show the added value of employing two complementary models.

A study of institutional aspects provided the insight that even though international institutions (such as the EU and the WTO) may provide both incentives and barriers for the introduction, marketing and promotion of NPFs, on balance the barriers exceed the incentives. Since these barriers (primarily intended to resist protectionist practices in international trade) cannot be circumvented, a successful introduction and marketing of NPFs should be taking them into proper account. In short, it is easier to start from already authorised foods and ingredients than to start from foods and ingredients that still need to be authorised, especially in Europe if they are GM crop derived. For the promotion of NPFs not too much should be expected from traditional government instruments such as taxes (on meat) and subsidies (on NPFs). Subsidies have already lost their appeal in most EU countries for purely domestic reasons, plus they are heavily restricted by EU regulations concerning state aid and the single market. If NPFs are to become a success, it should primarily be through private, commercial means and action. International institutions can protect and support commercial interests, however, through the international protection of intellectual property rights.

7.4.2 Further insights and tools required

Each of the tools has provided relevant information about opportunities and barriers for NPFs, but, under the present conditions of uncertainty, none could lead to conclusive answers. For example, one of the drawbacks of consumer research is that the currently available meat substitutes are, in fact, sold in a niche market and that they are almost twice as expensive as the cheapest meats. In order to improve the relevance of consumer research (sensory research, in particular), actual pea-derived NPF products – in the form of cheap protein products or as quality products – are indispensable. Since such products are not available yet, they should be crafted for this purpose. These products could also be used in assessing the relation between sensory properties of NPFs and their physical characteristics, a very difficult field of research, but of great importance for consumer-oriented product development. Evidently, it is important for product developers to realize that "the average consumer" does not exist. Subsequently, it is important to realize that different NPFs will have to be developed to fulfil the needs of different consumer groups.

All of the tools may seriously be influenced by the use of pre-conceptions in thinking about the future. The role of pre-conceptions is particularly great when questions have to be answered about the social desirability of policy options. It was shown that pre-conceptions lead to hidden assumptions that blur the arguments for or against an option. For instance, one of the pre-conceptions defined a diet shift primarily as an opportunity to develop products with a larger profit margin than meat. In contrast, it may be more sensible to develop plant protein ingredients that can serve as a low price alternative to meat protein ingredients. Another example is the implicit assumption mentioned above that meat prices will continue to decrease in the future, in consequence of continuing agricultural efficiency. Probably, policymakers may show wisdom by expecting that the meat prices will not decrease but increase as a result of changes in the world market and agro-ecological constraints on production. In fact, it seems inevitable that the present growth of spending power in China will put the world market prices of meat and feed under pressure. In addition, it should be emphasized that not too much should be expected of traditional government instruments such as taxes and subsidies. Although the latter are effective tools, their application is fraught with political difficulties.

Although the various tools to analyse the social desirability of a diet shift have not identified strong arguments against it, a critical reflection on the results may lead to the conclusion that the corresponding PROFETAS hypothesis deserves some more stringent tests than the arguments that have been described. In addition to the limitations mentioned above, it should be noted that the present analyses might have overstated the importance of agriculture to the political and economic processes in modern society. Other linkages, in particular to non-food issues, may have been overlooked or underestimated.

It is clear that health issues – which have not been directly studied in PROFETAS – require further attention. For example, it has emerged recently that *intensive* production of poultry and pigs in close proximity with people may play an important role in the adaptation of originally poultry-specific viruses via pigs to human beings as suitable hosts (Pilcher, 2004; Chen *et al.*, 2004). Under such conditions – primarily extant in South East Asia – new viruses are frequently spawned. It seems more and more likely that recent incidents such as with SARS and avian influenza are correlated with the intensive meat production there, and that the frequency of such incidents will continue to rise in parallel in the future. Apart from animal welfare issues, this health aspect, in itself, seems a good reason to reduce intensive meat production in general, and that of poultry and pigs in particular, at least under conditions such as in South East Asia, which is globally an important producing and exporting area.

7.5   TRANSITION FEASIBILITY

7.5.1   Insights delivered by PROFETAS

From PROFETAS the conclusion emerges that there are several sound reasons that support the triple hypothesis. That is, a shift in the Western diet from meat proteins to plant-derived protein products appears to be (1) environmentally more sustainable than present trends, (2) technologically feasible, and (3) socially desirable. Interestingly, the citizen generally seems to consider sustainability to be socially desirable, but the consumer does not like the taste of present meat substitutes. Nevertheless, the main evidence is in support of a transition to make food production and consumption more sustainable. Given the aims of PROFETAS, the programme has not included specialized transitions research. Nevertheless, many insights have been generated on the protein transition and on its linkages with the energy and freshwater transitions. These insights are described below.

First, it has become clear that the protein transition is a necessity at the global level. Without it, food production and sustainability are on a collision course. From time to time, there will be signals, alarming reports or dramatic events, which may contribute to the opening of a policy window. If the window opens, however, the market system will not simply guarantee that the most sustainable alternative technology will win. Market failures, but also government failures, may result in the selection of a suboptimal solution.

Second, study of the technological feasibility has revealed a number of options, but also some gaps in the required knowledge, such as the preferred crop for NPF production. Although many criteria for such a crop have been established (such as low fertiliser requirement, high protein yield, and preferably already part of the established food system), a more conclusive choice cannot be made yet, for criteria in other areas (such as concerning the non-protein fraction) have not yet fully materialised. Presently, the top choice in Europe may be rapeseed or pea yet, but in Asia it will probably be soy.

Third is the insight that the protein, energy and freshwater transitions are inextricably intertwined and that all parts of the crop or seed – protein and non-protein – should be considered one combined chain. Although the former, in particular, is a novel insight, both the former and the latter views fit in nicely with the present trend towards a biobased economy, aiming to derive food and feed, chemicals and other non-food materials, and energy from plant crops in the most efficient way.

Fourth, there is support from various actors, although this depends on linkages with other issues. Feedback from Dutch governmental actors and NGOs revealed enthusiasm for a protein transition, however, they are not inclined to initiate one themselves. In contrast, they do feel committed towards an energy transition. For most Western consumers, sustainability or the environment is not an incentive for food choice, however, health is. Due to a number of meat crises and other food scares, health and animal welfare are valued as increasingly important issues by Western consumers. In fact, sales of meat substitutes are increasing every year and they are being bought and eaten by non-vegetarian consumers.

In conclusion, several trends in different areas (protein, biofuel, water saving), on different geographic levels (local to global; western to developing countries) and concerning different actors (consumer, government, industry, NGOs) have been identified by PROFETAS, which – taken together – may lead towards a protein transition. A major step is raising the awareness that all these trends are linked up and cannot be seen in isolation. So a major achievement of PROFETAS is the insight to propose bringing together all these different actors, all with their own agendas and multiform aims. This is, we believe, the true definition of a societal transition.

7.5.2 Further insights and tools required

For pragmatic reasons, in PROFETAS many boundary conditions were assumed, such as the focus on the Western consumer and the focus on peas. At this stage, it seems timely to start focusing the attention on how – and where – a protein transition may be achieved in a dynamic, multiform society. Alternative options such as deriving proteins from algae should be investigated, as well as the feasibility of reducing the present protein overconsumption in Western countries (the "eating less protein" option). More direct research on transitions seems in order. The latter requires better insight into the role of explicit as well as implicit pre-conceptions in thinking about the future. In this respect, the use of multi-method strategies appears to be indispensable. Multi-method strategies may not only validate each other or specify complementary aspects of a complex phenomenon, but they may also show interesting divergences.

PROFETAS has shown several divergences that are extremely important for further research into transitions. Some examples are:

- The divergence between pre-conceptions that see a diet shift primarily as an opportunity to develop products with a larger profit margin than meat and pre-conceptions about developing plant based products that can serve as a low price alternative to meat protein ingredients.
- The contrast in the studies of consumer behaviour between, on the one hand, a series of impressive changes in dietary choices over the last few centuries and decades and, on the other hand, the observation that an individual will not easily change his or her food preferences from one day to the next.
- The assumption that meat prices will continue to decrease in the future, in consequence of continuing agricultural efficiency, versus the assumption that meat prices will increase as a result of changes in the world market and agro-ecological constraints on production. In spite of historical trends, it seems inevitable that the strong increase in meat consumption in countries such as China will put the world market prices of meat and feed under pressure.

Further research may be necessary into historical transitions, as well as into future general trends in global society. With respect to the latter, developing a number of contrasting visions of a potential global future is often considered helpful. In order to arrive at a robust transition strategy, these visions can be confronted with desirable options for protein food development. Such an approach should not be too general, for example, by focussing on aggregated parameters such as the growth rate of the world economy. More specific factors, such as global public health, should be an explicit part of the picture. So does taking geographic and cultural inhomogeneities into account, which clearly requires international cooperation.

7.6 GOVERNMENTAL POLICY OPTIONS

For the purpose of Dutch and European policy making, it can be argued that protein sustainability is a medium to long-term issue on a global scale. In the not too distant future, intensive meat production worldwide should be discontinued, or at least strongly reduced. In practice, the notion of an approaching collision between current food production and sustainability means that imminent disturbances of the protein production chain will let themselves be known to policymakers through all kinds of signals. Obviously, some of the weak signals could be a gradual rise of meat and feed prices. Stronger signals may include human and animal health incidents, such as with SARS and avian influenza, which are probably correlated with intensive meat production.

Against the background of irregular signals that vary in strength, governmental policymakers may have to manage at least three emerging goals:

- The first goal is to detect and interpret the signals correctly and to attribute them without delay to the relevant disturbances of the protein production chain. The BSE case has shown that a delay might have disastrous consequences on the controllability of the whole policy making process.
- The second goal is to minimize any negative effects that the disturbances may have on society. For instance, it may be wise to be prepared for a large-scale vaccination campaign.
- The third goal is to prevent the opportunity to take action getting lost as a result of counteracting processes that favour suboptimal solutions. That is, solutions that are suboptimal from the perspective of long-term sustainability objectives, not aimed at the root of the problem, such as temporary bans on certain types of meat.

The prevention of suboptimal solutions may require that governmental policymakers take action to correct market failures, for example, where private actors do not take full responsibility for the societal consequences of their activities. However, suboptimal solutions may also result from government failures, such as subsidies given to the wrong group or at the wrong moment. Moreover, the various linkages between the protein transition and other transitions on different levels of scale will seriously complicate any attempt to apply straightforward forms of strategic planning.

The least a robust policy should entail is an open mind to problem solving. For governmental policymakers it would be wise to avoid fixation on one particular technological option. In contrast, this might involve a decision to actively stimulate a *range* of divergent potential solutions to be developed, for example, by supporting research in the pre-competitive stage of technology development. This may especially be necessary where private parties have difficulties in making sense of the linkages between transitions, i.e. protein, energy and water. This approach would entail:
- Slighting the mental barriers between thinking about food, energy and water policy, or about natural resources in general.
- Trying to identify links with specific other policies, such as with the fight against obesity, the aims for renewable energy (such as the EU Biomass Action Plan) and the aims to promote organic farming.
- Considering alternatives such as NPFs as sustainable successors to animal products (such as pork) with regard to export of products and know-how and considering incentives to R&D in that area.

In combination with an open-minded approach, it is extremely important that governmental policymakers pay attention to the selection environment in which newly developed NPFs have to survive. Governmental policymakers can do a lot to prevent suboptimal solutions by monitoring and influencing the selection environment NPFs are facing. Some points in need of attention are the following:
- Initiating removal of national, EU and international (WTO) institutional barriers to the introduction of NPFs on the market.
- Increasing the transparency of the food chain in order to enable citizens and consumers to make sound choices.
- Continuing and internationally promoting "green" thinking by acknowledging the role of plants for improving sustainability.

In fact, the presently emerging trend towards a "biobased economy" is a first step toward the latter. In addition to deriving materials and energy from plant crops in combination, by the same token, food can be added to the list because crops are the ultimate renewables, degradable and all, and particularly sustainable if little fertiliser is required (such as with nitrogen-fixing protein crops).

European (EU) policymakers are in much the same boat as their Dutch counterparts, where they have to deal with global environmental changes. The goals for biofuels and organic farming are easily linked to the protein transition. In addition, recovering the presently lacking self-sufficiency in protein rich feed crops may be an incentive for EU policy towards striving for sustainability by means of a protein transition. This may be even more so in view of the expected increase in demands (and consequently, competition) for protein rich feed crops by industrialising countries such as China. An additional incentive may be provided by the potential savings on freshwater use by agriculture, given the rising need expressed by the World Water Forum.

7.7 INDUSTRIAL POLICY OPTIONS

For industrial policymakers an important question will be whether or not they want to be among the first movers in the market of new protein products. A company's decision on this topic will be governed by strategic and political circumstances at the time the options are contemplated, such as the ripeness of an issue. These circumstances depend on its own capabilities, its position in the industry, the economic situation of this industry and the industry's public image. The ripeness of an issue will in particular be influenced by the technological state-of-the-art (including innovative power), by the durability of the issue (a trend or a hype), by public opinions (including campaigns that emphasize the "ills" of an industry), and by linkages that improve the market success of a particular product segment.

In view of the many potential linkages between a protein transition and other transitions, it should be emphasized that developing alternatives to an established technology may require many innovations and that these will not occur automatically as the outcome of a linear process. The currently predominating policy of multinational companies seems to be rather risk-aversive by doing R&D themselves only on a moderate scale and, when deemed necessary, strategically buying emerging small companies with innovative products. The progress of small companies may be highly dependent on the relational context in which an innovation is located, such as the other innovations on which it builds, the status of the actors that own competing innovations, and the underlying community of experts involved in its elaboration. In addition, many chance factors may affect the content and the timing of innovation decisions.

Accordingly, the prospects of NPFs may depend on many initiatives. As a start, they may entail:
- Implementing "green" thinking in policies towards stepwise innovation and stepwise improving sustainability.
- Slighting the mental barriers between thinking about raw materials, energy and water. Integral thinking is a cornerstone of the emerging "biobased economy".
- Considering plant-derived alternatives as sustainable successors to animal products with regard to export of products and of know-how, and increasing R&D in that area.

This way of thinking coincides very well with parts of the mission statements of many companies: to create innovative products that promote a sustainable future.

Next to contributing to a sustainable future, NPFs could also meet another important issue for industry: making profit. Preliminary PROFETAS calculations indicate that cost price of NPFs could be lower or at least comparable to that of pork. Furthermore, meat prices are expected to increase. This increase will be higher than the increase in the cost of plant proteins because of the low conversion factor of plant protein to meat protein. Hence the gap between the costs of NPFs and meat will become bigger, thereby increasing the profit margin. Furthermore, NPFs are directed towards a growing market. Among others, the increase will evidently depend on the extent to which the products meet consumer demands and preferences, not only with respect to sensory properties but also to other issues such as health and animal welfare. This will be discussed in Section 7.8.

Those industrial policymakers who want to be among the first movers on the NPF market may start thinking about questions such as:

- Which markets are the most interesting options, specialty or bulk, which products and how to position the products on the market?
- What are the preferences of the different consumer groups and which ones can be translated to product characteristics?
- What techniques are required to produce the desired products?
- Which proteins or, even more basic, which crops seem feasible where?
- What chains should be developed and optimised, and with whom?

This sounds partly like revisiting technological projects of PROFETAS and indeed it is, but now on the level of actual and competitive products, rather than on the level PROFETAS was directed at, i.e. the development of a toolkit on a basic, pre-competitive level. However, to answer the questions asked above in a science-based way and to provide a sound base for future NPF development, further pre-competitive research on technological issues is still required. A number of subjects for such research have been discussed for example in Section 7.3.

In addition to technological issues, questions should be addressed such as:

- Cooperation with what other parties is useful (e.g. for optimal use of the crop)?
- How may a protein transition spread across the world (e.g. which countries are best suited for introducing and testing NPFs)?
- What may be the side effects of a protein transition (e.g. with respect to the North-South divide)?

- What trends in other areas (transport, technologies, lifestyles) may be relevant?

For smaller companies it may pay to closely watch the emerging trend towards a "biobased economy" and start thinking about strategic alliances. As long as innovation is not "directed" by larger companies (cf. transition "management" as proposed by the Dutch government), niches for innovative products will open, be it plant-derived protein foods or products made from the non-protein fraction of crops, products for saving water and/or energy, or others.

7.8 CONSUMERS AND OTHER STAKEHOLDERS

The consequences of a protein transition will affect many stakeholders, such as consumers, retailers, farmers and NGOs. Without addressing each of these groups individually, the present section intends to sketch the most pertinent implications. What consumers should ideally do in the context of a protein transition may boil down to (a) eating one third less protein (the average Dutch over-consumption), (b) replacing one third by plant-derived proteins, and (c) replacing the remaining third by extensively produced meat (such as most beef and lamb). Although in the Netherlands intensively produced pork converts much food industry waste in a sustainable way, globally this is not a representative example. Theoretically, the proposed threefold option has both environmental and health benefits. In practice, few consumers will be convinced immediately, but they may do so in the long run.

At present, the environmental benefits may not appeal to many consumers. The well-known activist storyline of "global nature" under threat and in need of protection from a global community has become too simplistic. Modern Westerners do not tend to think in terms of one big environment that is the same for everyone. They want credible solutions that give them the feeling that they are "doing the right thing." In contrast, the fact that many people are no longer aware of the animal origin of meat indicates that there is an increasing indifference toward the origins of proteins. At first sight, this seems to open possibilities for NPFs. Some producers might be tempted to change the protein chain in a way that does not have to be noticed by the people who are eating it, which may create a substantial shift from meat to plant protein foods without much consumer involvement.

On second thought, however, a low-involvement approach may not be the optimal strategy to pursue more sustainable food choices. If people are no longer aware of meat's animal origin, they will also be less inclined to pay attention to animal welfare. This may have negative consequences for attempts to stimulate sustainable agriculture by promoting high quality meat from well-treated animals or by simply eating less meat. An additional reason to not opt for a low-involvement approach refers to the societal value conflicts that are expected in many technology-related areas. In Europe, the recent example of genetically modified food has shown that a low-involvement approach may backfire if people get the impression that they are part of a "hidden" transition. Therefore, it is essential that all the people concerned are mindful of any transitions of food production methods.

One of the ways to involve people is a discussion on personal health aspects, which have many links to the protein transition. In contrast to plant-derived diets, meaty diets generally contain more saturated fats – associated with heart and coronary disease – and are sometimes associated with overconsumption of calories – leading to obesity. NPFs may not only exert a beneficial effect on health indirectly, via these relationships, but maybe also directly. That plant proteins may provide such an effect is indicated, for example, by the claim approved by the American Food and Drug Administration: "Diets low in saturated fat and cholesterol that include 25 grams of soy protein a day may reduce risk of heart disease" (FDA claim 21 CFR 101.82, October 1999). Proteins from crops other than soy seem to exert the same protective effect, according to the FDA, though this protective effect does not go unchallenged. Last but not least, NPFs are complex foods for which the amount of calories can be set via the choice of ingredients, which may be an important tool in view of the increase in obesity. Admittedly, certain plant-derived foods may generate more allergies than meat, but this affects a minority of people compared to the stifling incidence of heart and coronary disease, let alone the downright epidemic incidence of obesity (800 million people afflicted worldwide).

Public health aspects, such as food safety, add to personal health aspects perceived by consumers. Due to recent meat crises, both with chemical contaminants (such as dioxins, antibiotics, growth hormones) and pathogenic microbes (such as foot-and-mouth disease, avian influenza) European consumers are rather keen on food safety. Other aspects – such as the proposed relationship between intensive pork and poultry production in South East Asia and the increasing outbreaks of avian influenza – may not be topics consumers think about when they are shopping for food, but reminders of these issues may gradually induce "green" thinking by acknowledging the role of plants for improving their quality of life.

7.9 CONCLUSIONS

Environmental sustainability

The preceding chapters and sections have once again made it crystal-clear how large the environmental burden of meat production is. Conservatively estimated, it requires a 3-10 fold larger agricultural area and energy input and produces 3-10 fold more eutrophication than the production of plant protein. Not only does meat production bring about over 60 fold more acidification, but it also appropriates 30-40 fold more of the dwindling freshwater resources. In addition, it produces a lot more pollution by pesticides, heavy metals and antibiotics. As an entirely novel finding, PROFETAS has clearly demonstrated that the protein transition is coupled inseparably to two other societal transitions, namely those towards sustainable energy production and towards sustainable water use (Section 6.3). The freshwater link is evident from the resource difference indicated above. The energy production link primarily regards the release of agricultural area, which is freed for biomass production. Furthermore, the considerable proportion (60-80%) of non-protein biomass released as a by-product during NPF production may be utilised very efficiently for energy production. An evident case of win-win-win, with a triple environmental gain due to combined savings on protein, energy and water (Section 6.3).

Technological feasibility

Concerning technological feasibility many questions have been answered, both in primary production, processing and chain development. Generic as well as dedicated tools have been developed or are under development. Among the generic tools is the crop growth model (which still has to be validated) and the model for chain design (which is still under development). Applied knowledge concerning flavour-binding properties of pea proteins should be rated among the specialised tools. As might be expected, in addition to answers, many new questions have been raised as well (see Chapter 3). These concern, among others, the way in which process and ingredient selection may fulfil consumer wishes for NPFs within certain target groups (Sections 3.2, 3.3 and 6.4). This puts forward three major challenges: how to obtain a wide range of textures by using plant proteins, how these textures are affected by ingredients other than proteins and how to obtain the desired flavour? Next to questions regarding processing, also questions with respect to primary production remain to be answered. These questions concern among others the extent to which protein composition may be manipulated without affecting the viability of the plant in the field, and which crops are the most suitable to yield raw materials with the desired specifications (Section 6.2). Another major issue is the extent to which new developments in molecular biology and breeding may affect the suitability of crops for NPF production.

Societal desirability

From the sustainability perspective, the societal desirability has been established beyond doubt, particularly concerning resource and pollution aspects (land, water and energy uses, and their implications for biodiversity loss and climate change). Furthermore, increased availability of plant protein based foods will undoubtedly make an important contribution to food protein security, which is presently under increasing pressure from world-wide increasing meat consumption. Doubtlessly, such a transition will have a significant impact on North-South relationships and the poverty issue in the world (Section 6.5). Concerning societal desirability it has also become apparent that different actors can hold different interests and opinions here (Section 6.4). It has been clearly established that the consumer is the player who holds the key to a short-term protein transition and that Western consumers currently rate health above sustainability. For consumer-oriented product development, much more interaction is necessary between product developers and various user groups, such as consumers leaning towards health, convenience or culinary traditions.

Developing countries

Although PROFETAS was originally designed from the Western perspective, it has become clear that in developing countries the incentives to achieve a protein transition are not just different (Section 6.6), but generally much stronger (Section 6.5). In China, for example, meat production generates pressure by its inability to meet the national demand on top of the pressure generated by severe local pollution. Crops tailored to climatic (Section 6.2) and cultural (Section 6.6) characteristics are available. It can be concluded that, although the developed countries are primarily responsible for the unbalanced meat consumption referred to earlier, it is primarily the developing countries that are confronted with the effects (see Section 2.5). The latter, therefore, may experience stronger and more direct incentives to strive for a protein transition. However, the required technological expertise may be available in developed countries mainly. At any rate, if mineral oil and meat prices continue to rise, particularly in developing countries there will be opportunities for the onset of a combined protein plus biomass transition. Since the majority of the developing countries are particularly short of freshwater resources, the additional implicit water conservation will be considered an important bonus.

Facilitating transitions

In PROFETAS, neither transitions, nor their feasibility have been studied directly. Rather, insights and tools have been developed to facilitate a potential transition from meat protein to plant-derived protein products in the near future. These have been summarised in the present chapter. For the reasons outlined above we feel that a protein transition – reducing intensive livestock farming and cultivation of feed crops – is not just beneficial to the environment, but also more sustainable, most certainly socially desirable, and in the long run inevitable. It is yet unclear whether NPFs will be able to replace meat, to what extent, on what scale internationally, and how rapidly. When the window of opportunity opens, however, it will be crucial to have the technology available for more sustainable alternatives.

Facilitating a transition does not seem easy. However, it is interesting to note that – while exclusively working on an approach to make protein production and consumption more sustainable – the PROFETAS research community has yielded an integrated solution for various global problems far beyond its original scope. In addition to promising much more sustainable protein production – directly contributing to biodiversity and resource conservation and probably indirectly to animal welfare and human health – PROFETAS simultaneously indicated realistic options to produce a significant amount of biomass for sustainable energy production, and to save an immense volume of freshwater. This suggests that integral thinking combined with an even further extension of the disciplines involved (including health aspects, in particular) may be a promising avenue to meet the foreseeable challenges of the next few decades.

REFERENCES

Chen, H., Deng, G., Li, Z., Tian, G., Li, Y., Jiao, P., Zhang, L., Liu, Z., Webster, R.G., and Yu, K. (2004), "The evolution of H5N1 influenza viruses in ducks in southern China", *Proceedings of the National Academy of Sciences*, Vol. 101 No. 28, pp. 10452-10457.

Pilcher, H. (2004), "Increasing virulence of bird flu threatens mammals", *Nature*, Vol. 430, p. 4.

HISTORY OF THE FUTURE (EPILOGUE)

Maybe you would like to know how a research programme on proteins came to yield solutions to many different kinds of pressing global issues as diverse as energy, freshwater, biodiversity and health ...

In 1995, the Netherlands Organisation for Scientific Research NWO – the Dutch equivalent of the British Research Council and the American National Science Foundation – laid the foundations for an innovative multidisciplinary research programme called "Knowledge Enriches" (Kennis Verrijkt). Under this initiative, NWO invited the – rather unique – combination of the Research Schools SENSE (Socio-Economic and Natural Sciences of the Environment) and VLAG (Nutrition, Food Technology, Agrobiotechnology and Health Sciences) to develop a joint research programme devoted to "making food production more sustainable", expected to address technology in a societal context. Among the initiators were Hans Opschoor, Paul Berendsen, Leo Jansen, Hans Linsen, Rob Hoppe, Wim Jongen, Klaas van 't Riet and Pier Vellinga. A steering committee was formed, which soon came to the conclusion that the largest environmental gain was to be anticipated in the protein chain.

For environmental and social sciences, the foundation was laid by transitions research and the International Human Dimensions Programme on Global Environmental Change. For food technology, a pillar was the "DTO" exercise (see Chapter 1) on Sustainable Technology Development, which included a case on NPFs, quoted throughout the book. The common ground was found in a "WRR" report, written by an advisory body to the Dutch government, entitled "Sustainable risks are here to stay" (Duurzame risico's: Een blijvend gegeven). A chapter of the latter report described two extreme interpretations of sustainable global agriculture, (a) globally concentrated, intensive (high input) agriculture on optimal soils in optimal climates (for example, cereals in the American corn belt) in tandem with subsequent transport across the whole world and (b) diffuse, local extensive (low input) agriculture for local use. The former is characterised by high yields on a limited area, however, with high inputs of fertiliser, water, pesticides and energy (for ample farming equipment, refrigerated storage and global transport). The latter is characterised by relatively low yields, high requirements of lower quality soils, low inputs of fertiliser, pesticides and energy (low transport), but high irrigation water requirements. The two extremes were sketched in an attempt to get more insight into food sustainability, and this approach raised many questions for further research.

In 1997, a scoping study was started including bilateral meetings with representatives of interested research establishments, ministries and industry. In an interactive process, the programme slowly matured. Potential participants suggested research topics for constituent projects. In a common workshop preliminary proposals were discussed among researchers, and in a second workshop with representatives of NWO, government, industry and NGOs. On the basis of the comments, the proposals were adapted and detailed. After extensive internal and external review, the complete set of concerted proposals was approved by NWO. In May 1999, the programme was commissioned and PROFETAS was set to go.

Enlisting PhD students proved harder than anticipated due to the job market, on which it was particularly difficult to find food technologists, economists and people experienced in multidisciplinary research. A full year was needed to complete the first contingent of nine PhD students. Subsequently, postdocs were hired and a process of shaping a concerted multidisciplinary research community was initiated. A plan for internal and external multidisciplinary communication was devised, with a primary role for two-day common meetings in remote places, in which sometimes social scientists were asked to relay their natural scientists' results and vice versa, in the presence of supervisors and Programme Committee members. By mid 2000, the PROFETAS research community had grown to about fifty people.

Clearly, PROFETAS was an experiment in multidisciplinary cooperation, as had been intended. In hindsight, the need (and the required funds) for extensive communication of a research community of this diversity (ranging from economics to food technology, from consumer research to ecology, and from food chemistry to political science) and this size had been severely underestimated both by the initiators and by NWO. By flexible reallocation of existing funds, however, and by acquisition of additional funds, this potential problem was largely overcome. It was not easy, but with a lot of devotion and effort from everyone concerned, the job was done just right. After a positive mid-term review in 2002, the programme came into full swing. The first PhD student papers were published and postdocs started to integrate monodisciplinary PhD student results. Some timely adjustments were made, notably, adding one project on non-protein fractions. Subsequently, the programme shifted gear once more, the integrated results became more and more interesting, to the point that – much to our own surprise – they started to transcend the scope of the original goal. As described in Chapters 6 and 7, the unbreakable links between the protein, water and energy transitions unexpectedly emerged from the fog of research.

Though the PROFETAS goal was focussed on the protein chain exclusively as a way to reduce the environmental pressure of food production stepwise, instead of gradually, it transpired that, completely unsolicited and apart from environmental benefits, PROFETAS may contribute to many other pressing global problems, including health, energy, freshwater and biodiversity. Without a truly multidisciplinary approach, we feel this would not have been possible. Therefore, we would like to conclude this book by expressing our heart-felt appreciation to everyone involved, sponsors, researchers, supervisors, Programme Committee members and interviewees alike, as well as with a strong recommendation to funding bodies all over the world to note the unexpectedly high yield of truly interdisciplinary results (stemming from multidisciplinary cooperation), that may hold promise for the future. May knowledge enrich and may multidisciplinarity multiply.

INDEX

α-subunits 63-65
2S albumins 164
acidification 31, 43, 44, 46, 175, 196, 213
actors 7, 32, 79, 129, 133, 134, 156, 175, 177, 181, 183, 184, 186, 196, 205, 207, 209, 214
Africa 165
AGE model 32
aggregation 10, 26, 47, 53, 64, 66, 78
agricultural markets 138, 150
air classification 27, 55, 87, 89, 91, 172
air pollutants 29
albumins 59, 61
aldehydes 53, 55, 65, 73
animal origin 103, 105, 108-110, 168, 211
animal welfare 4, 23, 24, 33, 101, 103, 107, 109, 110, 113, 135, 147, 177, 184, 188, 204, 205, 210, 212, 215
antibiotics 167, 168, 212, 213
appearance 62, 113, 115, 116, 119, 121-123
appetite for meat 4
applied general equilibrium (AGE) model 32, 139, 143
appropriateness 102, 117, 120, 123
Asia 32, 34, 47, 52, 165, 188, 195, 196, 204, 205, 212
Asian consumer 165, 188
Australia 165
authorisation 129, 145, 149
avian influenza 4, 204, 207, 212

backcasting 7
bakery products 157
barriers 7, 88, 112, 113, 136, 143, 145, 146, 149, 150, 183, 202, 203, 208, 209
beany flavour 58
beef 13-15, 24, 105, 106, 108, 146, 173, 174, 211
behaviour 47, 61-63, 65, 66, 74, 78, 93, 95, 99-105, 109, 111, 123, 124, 144, 161-164, 175, 176, 183, 197, 198, 200, 201
behaviour of consumers 56, 57, 99, 175, 176, 177, 200
biobased 88
biobased economy 193, 205, 208, 209, 211
biocrude 88
biodiesel 88, 89, 164, 170
biodiversity 3, 4, 24, 25, 35-39, 41, 171, 179, 185, 186, 196, 214, 215, 217, 219

bioethanol 164
biofuels 37, 164, 170, 173, 198, 205, 208
biomass 36, 37, 53, 70, 72, 73, 156, 168, 172-174, 180, 199, 208, 213, 214, 215
biorefinery 170
bite 56
bitterness 58, 61, 121, 198
Black Death 104
Brazil 31, 155, 182, 194
Brundtland commission 2
BSE 4, 111-113, 147, 148, 168, 207
bud-containing tissue (BCT) 76, 77
bulk 203, 206, 207
business strategies 176, 177
by-products 28, 29, 91, 131, 132, 135, 136, 139, 155, 157, 166-169

cake 172, 188
campylobacter 4
Canada 75, 165
canopy nitrogen 68
carbohydrates 27, 51, 52, 55, 86, 93, 161
carbon cycle 36
causal networks 26, 42, 47, 196
cellulose 86, 87, 167
chain design 79, 80, 82, 85, 94, 198, 200, 213
cheap bulk proteins 180, 201
cheese 13, 52, 62, 118
chick peas 43, 165
China 155, 182, 188, 189, 194, 203, 206, 208, 214
citizen-consumer 34, 183
citizens 23, 33, 130, 133, 196, 208
Classic 69, 76
coagulated textures 62
co-firing 88
common agricultural policy (CAP) 142
concentrates 52, 55, 83, 138, 157, 160, 161, 163
consumer research 118, 200, 203, 218
consumer segments 62, 111, 179
consumer-driven approach 117
consumerism 104, 105, 137
consumers 2-5, 7, 8, 17, 27, 32-34, 43, 47, 56-59, 61, 62, 66, 79, 82, 83, 91-95, 99, 100, 103, 106-111, 113, 115-121, 124, 130, 135-137, 139, 140, 144,

145, 147-149, 151, 160, 165, 167, 173-180, 183-188, 193, 196, 198-201, 203-206, 208, 210, 212-214
consumption 1, 4, 7-12, 17-19, 25, 27-30, 32, 33, 39, 40, 43, 45-48, 57, 58, 61, 91, 94, 103, 105, 107-116, 120, 132-134, 136, 138-140, 142, 143, 147-149, 151, 163, 167, 169, 170, 172, 173, 179, 181-186, 188, 189, 193-196, 201, 202, 204, 206, 214, 215
consumption patterns 12, 48, 138, 174, 196
consumption studies 123
conventional environmental issues 31, 35, 36
conversion 4, 29, 36, 41, 45, 51, 83, 134, 169, 170, 172
conversion factors 4, 26, 132, 134, 135, 210
conversion ratio 31, 37
convicilin 55, 63, 65, 74, 75, 78
cost 11, 79, 81-83, 90, 134, 176
cost calculations 85, 94
cost price 81, 94, 198, 210
crop choice 155, 156, 162, 164, 165, 200
crop growth model 68, 69, 72, 92, 94, 162, 197, 199, 213
crop yield 67, 68
crop-based solutions 155, 156, 187
cropland 37, 141, 151, 171, 173, 175, 176
crust 119
cultivars 65, 66, 69, 70, 72, 74, 92, 93
   Athos 69
   Attika 69
   Classic 69, 76
   Espace 65, 69, 76
   Finale 65
   Phonix 69
   Puget 76
   Santana 69
   Solara 60, 65, 69, 70, 76
   Supra 60, 65, 76
cultivation in Europe 159
cultivation systems 92, 93

denaturation 64
desserts 157
developed countries 34, 134-136, 155, 194, 214
developing countries 44-46, 48, 134, 135, 137, 141, 182, 188, 194, 196, 205, 214
dietary choice 105, 124, 200, 206
dietary patterns 101, 102, 110
diffusion 142-144
dilemma 183
dish 106, 107, 117-124, 201

distribution 3, 42, 79, 83, 141, 142, 144
disulfide bond 64, 65
dressings 157
drought 73, 164
dry peas 83
dry process 55

ecological footprint 25
econometric modelling 131, 132
economy 2, 18, 26, 32, 47, 107, 133, 139, 156, 186, 187, 193, 195, 205, 206, 208, 209, 211
ecosystem functions 34-36, 39, 41, 42, 45, 46
ecosystem services 26, 34-36, 47, 195
egg albumin 53
eggs 10, 13, 14, 118
elasticity 63, 122
emerging options 193
energy 29, 35-38, 43-45, 53, 54, 80, 83, 87-90, 94, 114, 151, 155, 156, 164, 165, 167, 174, 176, 182, 185, 186, 199, 204, 205, 208, 209, 211, 213, 215, 217-219
energy use 3, 29, 37, 40, 43, 81, 173, 175, 186, 214
Engel curve 139
engineering culture 104
environmental load 73, 81-83, 85, 93, 94, 198
environmental standards 33, 34, 47, 195, 196
Espace 65, 69, 76
ethanol 170
eutrophication 4, 31, 35, 41, 43, 44, 46, 156, 196, 213
exergy 81-85, 94, 198
extending 136
extrusion 52, 53, 66, 83, 163, 199

fair trade 23-25
farmers 48, 135, 158, 159, 164, 1167, 199
feed area 172
feed 27-29, 31, 32, 43, 46, 48, 53, 54, 83, 86-91, 94, 132, 134-141, 145, 155-157, 160, 166-172, 174, 182, 188, 195, 196, 198, 203, 205-208, 215
fertilisation 35, 187
fertilisers 29, 31, 37, 38, 43, 44, 54, 69, 164, 167-169, 175, 182, 205, 208, 217
fibre-like textures 62
fibrous 8, 53, 122
Finale 65
firmness 63
fish 11, 13, 14, 24, 34, 106, 108, 118
fixing nitrogen 54

flavour 56-62, 81, 83, 92, 93, 117-123, 156, 162, 197-199, 201, 213
flowering 73
focus group 118-120, 123
food politics 138
food safety 24, 180, 187, 188, 212
food scares 5, 176, 205
food security 10, 24, 188
food supply chain 79, 80, 83, 100
food sustainability 11, 23, 151, 217
food system 1, 3, 29, 51, 61, 138, 150, 151, 161, 181, 194, 202, 205
foot-and-mouth disease 4, 111, 212
forest 37, 171, 174
fossil fuels 35
fowl 106
freed-up area 173
freshwater 4, 24, 182, 186, 204, 205, 213-215, 217, 219
freshwater availability 3, 43, 194
freshwater use 46, 156, 175, 186, 209
fuel 88
fungal diseases 73
Fusarium 5, 53, 73
future meat price development 14, 41, 202

general agreement on tariffs and trade (GATT) 146
genetic modification 74, 76, 78, 118, 162, 164, 177, 180, 197
genetically modified organisms (GMOs) 145, 146, 149
genotype 67-69, 73, 76, 78, 197
geochemical cycles 48
global food economy 156, 187
global starch production 163, 170
global warming 31, 41, 43, 44
globalisation 12, 131, 182
globular 8, 59
globulins 55, 56, 73-75, 78
grass 86, 158-161, 199
grazing 183
greenhouse gases 29, 35
gross domestic product (GDP) 15, 17, 142
groundnuts 54, 165

harvesting 53, 73
health claims 185
health issues 204
heat-induced gelling behaviour 62, 63, 93, 95, 197
heavy users 112, 115
herbicides 29
higher heating value 173
high-quality specialty products 180
hormones 76, 135, 146, 212

human health 4, 23, 24, 41, 135, 136, 146, 180, 187, 215
hydrological cycle 48

ice-free land 3, 43
ideals 3, 24, 180
ills 3, 24, 180, 209
incentives 5, 103, 142, 145, 149, 176, 194, 202, 205, 208, 209, 214
India 165
indicators 12, 24, 26, 29, 31, 40-45, 47, 48, 80-82, 195, 196
industrialized countries 44, 45
ingredient preparation 79, 83, 118
ingredients 4, 8, 12, 27, 43, 53, 83, 84, 93, 107, 115, 117, 118, 121, 145, 149-151, 157, 167, 201-203, 206, 212, 213
innovations 5, 109, 131, 177, 180, 209
institutions 129
integrated crop management 23
intellectual property rights 146, 147, 149, 150, 202
intensive livestock farming 199, 215
intensive livestock sector 8, 193
intensive meat production 174, 198, 204, 207
intensive production 31, 204
intercontinental cooperation 189
international institutions 131, 132, 144, 145, 149, 150, 201, 202
international trade 25, 54, 138, 145-150, 202
irrigation 35, 37, 70, 217
isolates 55, 56, 63, 65, 157, 160, 161, 163

juiciness 156, 121
just-in-time production 183

ketones 59, 61

label 101, 113, 114, 148
land 3, 24, 28, 29, 31, 35, 37, 79, 90, 141, 144, 155, 156, 167-169, 171-175, 182, 186, 199, 202, 214
land use 26, 31, 35, 39, 79, 90, 138, 139, 141, 167, 171-173, 184, 196
landscape quality 70, 71
leaf area index 68, 69
leaf photosynthesis 68
leaf senescence 68, 72
legumin 55, 59, 61, 63-65, 68, 74, 75, 77, 78, 163, 164
lentils 54
life cycle analysis (LCA) 8, 25, 30, 40, 45
life cycle assessment 87
lifestyle 103, 105, 111, 112, 140, 141, 188, 211

livestock 4, 8, 32, 35, 139, 172, 193, 199, 215
livestock feed 188
lodging 73, 164, 197
lucerne 5, 158-161
lupine 53, 157-161

macroeconomic 5, 132, 138, 141-143
manure 28, 31, 37, 38, 44, 85, 168
margarine 7
market failures 132, 205, 207
market power 177
market segmentation 111, 178
market system 7, 130, 204
meals 27, 52, 102, 103, 107-110, 114-117, 119, 123, 157, 180
meat chain 10, 25, 84, 85, 166-168, 181, 194
meat crises 4, 111, 205, 212
meat extender 52
meat prices 132, 144, 150, 151, 180, 201-203, 206, 210, 214
meat substitutes 5, 8, 52, 53, 56, 91, 92, 110-124, 132, 143, 157, 165, 176, 183, 185, 200, 201, 203-205
meat substitutes market 56, 110, 112
meat's animal origin 103, 105, 108-110, 211
Mediterranean 15, 17, 19, 110
micro-organisms 5, 51, 145
milk 10, 13-15, 19, 59, 142
monogastrics 139
monopolistic behaviour 144
multidisciplinary 8-10, 193, 217-219
multi-method strategies 25, 195, 206
mutton 13, 14, 174
mycoprotein 53, 115, 118, 121, 136

natural area 35, 37, 38
natural resources 4, 23, 24, 44, 155, 186, 208
natural variation 74, 78, 92
nature conservation 173
near wilt 73
neophobia 112, 113, 117, 201
N-fixation 68, 69
NGOs 193
niche market 124, 166, 170, 173, 203
nitrogen cycle 36
non-protein fraction 25, 29, 40, 57, 85-91, 95, 170, 197, 198, 200, 205, 211, 218
non-starch polysaccharides (NSP) 86-88, 91, 170
non-tariff barriers 146
non-vegetarian consumers 111, 116, 124, 201, 205

Novel Food Regulation 145, 149, 184
Novel Protein Foods (NPFs) 2, 5-8, 27, 29, 30, 32, 33, 34, 37-39, 43-45, 47, 51, 52, 56-59, 61, 62, 65-67, 74, 79, 82, 83, 85, 86, 90-95, 97, 101, 102, 109-111, 113-118, 123-125, 130, 131, 137-140, 142-144, 146-151, 156-158, 161-166, 168-172, 175-181, 183-188, 193, 195, 196, 203, 205, 208-215, 217
NPF chain 25-31, 43, 46, 82-85, 94, 166, 194, 197, 198
nutrients 29
nuts 13, 118

odour 58, 122
offal 13, 14, 135
off-flavours 58, 59, 92, 93, 198, 199
oil 13, 28, 34, 46, 54, 57, 83, 86, 88-91, 94, 136, 157, 158, 163-165, 167-173, 176, 188, 198, 200, 214
organic agriculture 23, 177
organic farming 208
organic food 113
organic meat 148, 183
over-consumption of protein 19, 211
ozone depletion 41

package 113, 114, 116, 152, 155
packaging 83, 107, 116
pancakes 119
paper 86, 174, 218
path dependency 12, 17, 19
pea flour 28, 60
pea genotypes 67, 69, 78
pea protein characterisation 63
pea protein composition 65, 74, 78
pea protein extraction 55
pea protein fractionation 63
pea proteins 51-55, 58, 59, 61-66, 73, 74, 84, 87, 91-94, 121, 145, 162, 163, 166, 172, 197, 198, 213
pea soup 83, 84
pea varieties 58-61, 197, 198
peas 5, 27, 28, 40, 43, 45, 54, 55, 57-61, 63, 69, 70, 72, 73, 83, 89-95, 163-165, 170, 173, 188, 197-200, 206
People 3, 5, 15, 23, 24, 39, 101, 104-107, 109, 110, 124, 132-134, 137, 141, 150, 201, 202, 204, 211, 212, 218
permanent pastures 171, 174
pesticides 3, 4, 29, 31, 164, 167, 182, 183, 213, 217
pet food 169
phosphate cycle 36

INDEX

pig feed 88, 89, 167, 168
Planet 94
plant breeders 78
policies 5, 9, 10, 41, 131-133, 146, 189, 208, 209
policy instruments 151
policy window 6, 179, 180, 194, 204
policymakers 6, 9, 10, 34, 136, 138, 156, 168, 179, 181, 182, 194-196, 203, 207, 208, 210
politicians 133
pollution 4, 31, 32, 34, 156, 186, 189, 213, 214
population growth 2, 3, 134, 138, 188
pork chain 26, 28, 29, 31, 43, 146, 48, 83, 84, 94, 166, 195, 196, 198
post-translational processing 78
potato 13, 14, 108, 157-160, 164
potato protein 160
poultry 13-15, 108, 172, 174, 204, 212
preferences of consumers 8, 56, 57, 59, 61, 91, 93, 95, 100, 109, 123, 154, 193, 200, 210
primary production 8, 25, 43, 45, 52, 56, 67, 79, 82, 92, 93, 107, 160, 162, 197, 213
problem owner 182, 186
processing 8, 9, 25, 27, 29, 37, 39, 40, 43, 47, 52, 55-60, 74, 75, 78, 79, 82, 83, 87, 90, 92, 93, 95, 101, 107, 130, 135, 136, 144, 148, 157, 160, 161, 163, 164, 168-170, 178, 179, 197-199, 213
producers 3, 4, 32, 47, 48, 99-101, 103, 107, 109, 110, 123, 124, 135, 137, 141, 142, 144, 147, 148, 164, 167, 168, 174, 184, 185, 188, 200, 211
product concepts 199
product processing 27, 29, 79, 83
production costs 79, 83, 136
PROFETAS philosophy 27
Profit 94
profit margin 150, 203, 206, 210
protein composition 55, 65, 73-75, 78, 92, 93, 197-199, 213
protein-flavour interactions 58
protein gelation 63-65
protein preparations 55, 58, 59, 61-63, 157, 161, 198
protein production 1, 8, 30, 36-40, 67, 68, 72, 138, 158, 159, 186, 193, 199, 202, 207, 215
protein quality 158, 159
protein supply 6-9, 12, 13, 16, 17, 19, 23, 182, 184-186, 193
protein yield per hectare 94, 197

protein, over-consumption 19, 211
protein-flavour interactions 57, 59, 162, 198
proteolytic processing 74, 75, 78
proximate decision makers 129, 130, 150, 151, 175
Puget 76
pulses 13, 14, 188

qualitative model 80, 81
quality 26, 31-33, 56, 58, 62, 67, 71, 72, 79, 81-83, 85, 91, 92, 94, 100, 110, 157-159, 161, 167, 176, 177, 179, 180, 187, 195, 197, 198, 201, 203, 212, 217
quantitative model 80
quinoa 158-160
Quorn ^{TM} 27, 53, 136

rapeseed 89, 90, 94, 157-160, 163, 165, 205
rate of cooling 93, 198
ready-made meals 107, 109
recommended daily intake 18
regional approaches 189
repeated consumption 110, 111, 115, 116
resource use 31, 40-44, 46, 48, 67, 68, 92, 197, 199
retail 9, 12, 18, 27-29, 39, 79, 83, 176
retail cost 11, 83
risks 7, 146, 217
root rot 73
ruminants 86, 139, 141

saponins 58, 60-62, 162, 198
SARS 204, 207
satiating properties 114, 116, 201
satiety 114, 115
scenario 39, 67, 70, 71, 82, 87, 140-144, 171, 173, 174, 200, 202
scorecard 88
seed protein production 67, 68, 70, 72
selection environment 129, 131, 150, 201, 208
sensory panel 117, 121, 122
sensory preferences 102, 118, 123, 201
sensory properties 113-116, 203, 210
sensory quality 56, 91
single cell protein 53, 136
slaughter 29, 106, 169
smell 58, 113, 115, 121
social status 101
societal transitions 1, 187, 205, 213
socio-cultural changes 101, 103, 104, 123, 200
soil nutrients 29, 43
soil quality 71
Solara 60, 65, 69, 70, 76

soups 8, 83, 84, 119, 121, 157
South America 34, 47, 195, 196
soy 8, 27, 28, 52-54, 57, 63, 64, 66, 74, 78, 89, 90, 94, 95, 114, 121, 136, 157, 162-165, 167, 170, 172, 173, 188, 189, 199, 205
soy concentrates 52
soy meal 52, 136
soy oil 28, 46, 170, 198, 200
soy proteins 52, 53, 58, 59, 65, 66, 118, 121, 136, 157, 162, 163, 165, 182, 198, 199, 212
specialties 184
species 5, 24, 37, 41, 54, 74, 76, 78, 105, 161, 168
Spirulina 5, 51, 52
stakeholders 11, 24, 131, 150, 156, 193, 199, 201, 211
Starch 28, 53-55, 57, 66, 73, 85-91, 94, 157, 158, 163-165, 167, 169, 170, 172, 173, 198
stock 88
straw stiffness 73, 164
subsidies 132, 147-150, 170, 202, 203, 207
sulphydryl 64, 65
supermarket 12, 107, 108, 113, 114
supply chain 9, 27, 79-83, 85, 94, 95, 100, 107, 182, 187
sustainable cropping systems 73
sustainable development 2, 23, 24
Sustainable Technology Development (STD) 7, 51, 217
swill 168
symbiosis 54
syngas 88

tapioca 27, 28
taste 5, 54, 56, 58, 60, 61, 79, 81, 89, 93, 100, 101, 102, 107, 113, 115, 117, 119, 121, 122, 164, 178, 198, 204
tax 5, 147-150, 202, 203
technology development 131, 201, 208
tempeh 5, 52, 165, 188
texture 5, 53, 56, 57, 62, 66, 79, 81, 92, 93, 113, 118, 119, 121, 122, 198, 199, 213
texture formation 57, 62, 63, 93, 156, 162, 1633, 197
texturisation 53, 162, 199

texturised vegetable protein (TVP) 8, 52, 53, 132, 136, 137, 151
texturising 163, 164, 199
three components meals 108, 109
tofu 4, 52, 62, 118, 121, 165, 188
toolbox 117, 165, 193
toolkit 8, 210
toughness 119, 121, 122
trade-offs 32-34, 47, 194, 196
transition management 1, 7, 151, 194, 211
triglycerides 169
triple hypothesis 9, 193, 204
triticale 158-160

vegetable oil 91, 167, 169, 170, 173
vegetarian 108, 111-114, 185, 188, 189, 201
vicilin 55, 59, 63, 64, 66, 74, 75, 78, 163, 165
vicilin/convicilin ratio 75
viruses 204

waste 8, 24, 25, 38, 39, 43, 81, 84, 85, 87, 91, 94, 136, 168, 211
water 3, 28, 29, 31-39, 43, 55, 67, 68, 70, 73, 81, 84, 87, 155, 156, 161, 162, 165, 167, 169, 173-175, 183, 186, 196, 197, 205, 208, 209, 211, 213, 214, 217, 218
water cycle 35, 36
water holding capacity 81
water supply 35-38, 70
water supply scenarios 67, 70, 71
water use 331, 37, 39, 44, 169, 173, 174, 213
wet process 55
wet processing 55, 56, 87, 90
wheat gluten 53, 157, 165
willingness to pay 187
window of opportunity 1, 5, 6, 129, 176, 180, 215
woodlands 171, 172
world energy demand 199
world land 171
world population 1, 3, 4, 7, 155
world trade organization (WTO) 142, 144-147, 149, 181, 202, 208
World War II 7, 10, 11, 17

ENVIRONMENT & POLICY

1. Dutch Committee for Long-Term Environmental Policy: *The Environment: Towards a Sustainable Future.* 1994   ISBN 0-7923-2655-5; Pb 0-7923-2656-3
2. O. Kuik, P. Peters and N. Schrijver (eds.): *Joint Implementation to Curb Climate Change. Legal and Economic Aspects.* 1994   ISBN 0-7923-2825-6
3. C.J. Jepma (ed.): *The Feasibility of Joint Implementation.* 1995
ISBN 0-7923-3426-4
4. F.J. Dietz, H.R.J. Vollebergh and J.L. de Vries (eds.): *Environment, Incentives and the Common Market.* 1995   ISBN 0-7923-3602-X
5. J.F.Th. Schoute, P.A. Finke, F.R. Veeneklaas and H.P. Wolfert (eds.): *Scenario Studies for the Rural Environment.* 1995   ISBN 0-7923-3748-4
6. R.E. Munn, J.W.M. la Rivière and N. van Lookeren Campagne: *Policy Making in an Era of Global Environmental Change.* 1996   ISBN 0-7923-3872-3
7. F. Oosterhuis, F. Rubik and G. Scholl: *Product Policy in Europe: New Environmental Perspectives.* 1996   ISBN 0-7923-4078-7
8. J. Gupta: *The Climate Change Convention and Developing Countries: From Conflict to Consensus?* 1997   ISBN 0-7923-4577-0
9. M. Rolén, H. Sjöberg and U. Svedin (eds.): *International Governance on Environmental Issues.* 1997   ISBN 0-7923-4701-3
10. M.A. Ridley: *Lowering the Cost of Emission Reduction: Joint Implementation in the Framework Convention on Climate Change.* 1998   ISBN 0-7923-4914-8
11. G.J.I. Schrama (ed.): *Drinking Water Supply and Agricultural Pollution. Preventive Action by the Water Supply Sector in the European Union and the United States.* 1998   ISBN 0-7923-5104-5
12. P. Glasbergen: *Co-operative Environmental Governance: Public-Private Agreements as a Policy Strategy.* 1998   ISBN 0-7923-5148-7; Pb 0-7923-5149-5
13. P. Vellinga, F. Berkhout and J. Gupta (eds.): *Managing a Material World.* Perspectives in Industrial Ecology. 1998   ISBN 0-7923-5153-3; Pb 0-7923-5206-8
14. F.H.J.M. Coenen, D. Huitema and L.J. O'Toole, Jr. (eds.): *Participation and the Quality of Environmental Decision Making.* 1998   ISBN 0-7923-5264-5
15. D.M. Pugh and J.V. Tarazona (eds.): *Regulation for Chemical Safety in Europe: Analysis, Comment and Criticism.* 1998   ISBN 0-7923-5269-6
16. W. Østreng (ed.): *National Security and International Environmental Cooperation in the Arctic – the Case of the Northern Sea Route.* 1999   ISBN 0-7923-5528-8
17. S.V. Meijerink: *Conflict and Cooperation on the Scheldt River Basin.* A Case Study of Decision Making on International Scheldt Issues between 1967 and 1997. 1999
ISBN 0-7923-5650-0
18. M.A. Mohamed Salih: *Environmental Politics and Liberation in Contemporary Africa.* 1999   ISBN 0-7923-5650-0
19. C.J. Jepma and W. van der Gaast (eds.): *On the Compatibility of Flexible Instruments.* 1999   ISBN 0-7923-5728-0
20. M. Andersson: *Change and Continuity in Poland's Environmental Policy.* 1999
ISBN 0-7923-6051-6

ENVIRONMENT & POLICY

21. W. Kägi: *Economics of Climate Change: The Contribution of Forestry Projects.* 2000
    ISBN 0-7923-6103-2
22. E. van der Voet, J.B. Guinée and H.A.U. de Haes (eds.): *Heavy Metals: A Problem Solved?* Methods and Models to Evaluate Policy Strategies for Heavy Metals. 2000
    ISBN 0-7923-6192-X
23. G. Hønneland: *Coercive and Discursive Compliance Mechanisms in the Management of Natural Resourses.* A Case Study from the Barents Sea Fisheries. 2000
    ISBN 0-7923-6243-8
24. J. van Tatenhove, B. Arts and P. Leroy (eds.): *Political Modernisation and the Environments.* The Renewal of Environmental Policy Arrangements. 2000
    ISBN 0-7923-6312-4
25. G.K. Rosendal: *The Convention on Biological Diversity and Developing Countries.* 2000
    ISBN 0-7923-6375-2
26. G.H. Vonkeman (ed.): *Sustainable Development of European Cities and Regions.* 2000
    ISBN 0-7923-6423-6
27. J. Gupta and M. Grubb (eds.): *Climate Change and European Leadership.* A Sustainable Role for Europe? 2000
    ISBN 0-7923-6466-X
28. D. Vidas (ed.): *Implementing the Environmental Protection Regime for the Antarctic.* 2000
    ISBN 0-7923-6609-3; Pb 0-7923-6610-7
29. K. Eder and M. Kousis (eds.): *Environmental Politics in Southern Europe: Actors, Institutions and Discourses in a Europeanizing Society.* 2000   ISBN 0-7923-6753-7
30. R. Schwarze: *Law and Economics of International Climate Change Policy.* 2001
    ISBN 0-7923-6800-2
31. M.J. Scoullos, G.H. Vonkeman, I. Thornton, and Z. Makuch: *Mercury - Cadmium-Lead: Handbook for Sustainable Heavy Metals Policy and Regulation.* 2001
    ISBN 1-4020-0224-6
32. G. Sundqvist: *The Bedrock of Opinion.* Science, Technology and Society in the Siting of High-Level Nuclear Waste. 2002
    ISBN 1-4020-0477-X
33. P.P.J. Driessen and P. Glasbergen (eds.): *Greening Society.* The Paradigm Shift in Dutch Environmental Politics. 2002
    ISBN 1-4020-0652-7
34. D. Huitema: *Hazardous Decisions.* Hazardous Waste Siting in the UK, The Netherlands and Canada. Institutions and Discourses. 2002
    ISBN 1-4020-0969-0
35. D. A. Fuchs: *An Institutional Basis for Environmental Stewardship: The Structure and Quality of Property Rights.* 2003
    ISBN 1-4020-1002-8
36. B. Chaytor and K.R. Gray (eds.): *International Environmental Law and Policy in Africa.* 2003
    ISBN 1-4020-1287-X
37. F.M. Brouwer, I. Heinz and T. Zabel (eds.): *Governance of Water-Related Conflicts in Agriculture.* New Directions in Agri-Environmental and Water Policies in the EU. 2003
    ISBN 1-4020-1553-4

ENVIRONMENT & POLICY

38. G.J.I. Schrama and S. Sedlacek (eds.): *Environmental and Technology Policy in Europe. Technological Innovation and Policy Integration.* 2003
    ISBN 1-4020-1583-6
39. A.J. Dietz, R. Ruben and A. Verhagen (eds.): *The Impact of Climate Change on Drylands.* With a focus on West Africa. 2004   ISBN 1-4020-1952-1
40. I. Kissling-Näf and S. Kuks (eds.): *The Evolution of National Water Regimes in Europe.* Transitions in Water Rights and Water Policies. 2004
    ISBN 1-4020-2483-5
41. H. Bressers and S. Kuks (eds.): *Integrated Governance and Water Basin Management.* Conditions for Regime Change and Sustainability. 2004   ISBN 1-4020-2481-9
42. T. Flüeler: *Decision Making for Complex Socio-Technical Systems.* Robustness from Lessons Learned in Long-term Radioactive Waste Governance. 2006
    ISBN 1-4020-3480-6
43. E. Croci (ed.): *The Handbook of Environmental Voluntary Agreements.* Design, Implementation and Evaluation Issues. 2005   ISBN 1-4020-3355-9
44. X. Olsthoorn and A.J. Wieczorek (eds.): *Understanding Industrial Transformation.* Views from Different Disciplines. 2006   ISBN 1-4020-3755-4
45. H. Aiking, J. de Boer and J. Vereijken (eds.): *Sustainable Protein Production and Consumption: Pigs or Peas?* 2006   ISBN 1-4020-4062-8

www.springer.com